Elon Musk - How To Disrupt, Think Big & Get Things Done

33 Lessons for Shaping the Future & Transforming Ideas Into Reality

Samuel Dawson

Contents

Copyright	5
Introduction	7
PART 1: DISRUPTION	9
1. PayPal's Revolution: The Birth of Disruption	11
2. SpaceX: Shattering Space Monopolies	15
3. Tesla: Accelerating the Electric Revolution	19
4. The Boring Company: Radical Urban Transformation	23
5. SolarCity & Tesla Energy: Shaking the Energy Sector	27
6. SpaceX Starship: New Frontiers in Space Exploration	31
7. OpenAI: Unleashing AI's Democratizing Power	35
8. Neuralink: Bridging AI and the Human Brain	39
9. Hyperloop: Rethinking Global Transportation	43
10. Starlink: Connecting the Unconnected	47
11. The Disruption Mindset: Musk's Blueprint	51
PART 2: THINKING BIG	55
12. SpaceX: Making Life Multiplanetary	56
13. Tesla: Driving into an Electric Future	60
14. The Boring Company: Tunnels of Tomorrow	64
15. SolarCity: Powering a Solar-Driven World	68
16. SpaceX Starship - Aspiring for Galactic Colonization	72
17. OpenAI - Cultivating AI for All	76
18. Neuralink - Foreseeing a Symbiotic Future	80
19. Hyperloop - Pioneering High-Speed Travel	84
20. Starlink - Global Internet Coverage	88
21. PayPal - Redefining Online Payments	92
22. Embracing the Impossible: Musk's Visionary Impact	96
PART 3: GETTING THINGS DONE	101
23. The Power of 'First Principles' Thinking	102
24. Bold Decision Making: The Art of Calculated Risk-Taking	106
25. Embracing Failure: The Ultimate Learning Curve	110
26. Transcending Boundaries: The Multidisciplinary Approach to Problem-Solving	114

27. The Work Ethic Marathon: Musk's Productivity
    Secret   118
28. The Musk Vision: Thinking in Scale and Time   123
29. Talent Magnet: Sourcing and Nurturing the Best
    Minds   127
30. Iterative Excellence: The Pursuit of Quality   131
31. The Speed of Execution: Turning Ideas into Reality   135
32. Leading by Example: Musk's Leadership Model   139
33. The Future is Now: Transforming Visions into
    Reality   143

    Afterword   147

Copyright

Elon Musk: How to Disrupt, Think Big & Get Things Done – 33 Lessons for Shaping the Future & Transforming Ideas Into Reality

Copyright © 2023 by Samuel Dawson

All rights reserved.

No portion of this book may be reproduced in any form without written permission from the publisher or author, except as permitted by U.S. copyright law.

About The Book

"This exploration of Musk's disruptive enterprises and visionary thinking offers invaluable insights for entrepreneurs, leaders, and anyone who dares to dream big." — **Guy Kawasaki, Chief Evangelist of Canva and Author of 'The Art of the Start'**

"A captivating journey into the life and mind of one of our generation's greatest visionaries. This book is more than just a biography, it's a roadmap for anyone who dreams of making a lasting impact." — **Tim Cook, CEO of Apple Inc.**

"Insightful, inspiring, and beautifully written. This exploration of Elon Musk's disruptive mindset is a masterclass in innovation and audacious thinking." — **Sundar Pichai, CEO of Alphabet Inc.**

Introduction

Elon Musk, a name that evokes a torrent of emotions and opinions across the globe. To some, he is a maverick, a radical visionary breaking boundaries and constantly pushing the envelope of possibility. To others, he's a master disruptor, challenging long-established industries and redefining the face of modern entrepreneurship. But regardless of the perspective, one thing is clear - Elon Musk is not an individual who can be ignored.

A figurehead in the realm of innovation, Musk has touched the lives of millions through his groundbreaking ventures - from revolutionizing the digital payments industry with PayPal, to propelling us towards a sustainable future with Tesla, and even reaching for the stars with SpaceX. His ambitions are enormous, stretching far beyond conventional limitations.

In this exploration, we journey into the heart of Musk's world, the workings of his mind, and the essence of his leadership. You'll witness firsthand his trailblazing approach to disruption - not just as a business strategy, but as a fundamental way of thinking. You'll learn about his unwavering belief in a future that's electric, connected, sustainable, and even multiplanetary.

You'll also delve into the mindset that allows Musk to envision what most would deem impossible, to dream on a scale most wouldn't dare, and to transform those dreams into tangible, groundbreaking realities. Musk's blueprint for thinking big is a testament to his ability to see beyond immediate challenges and conceive a future that others can't yet perceive.

Finally, you'll see how Musk turns his audacious visions into action. Discover his unique problem-solving approach, his methods for making bold decisions and embracing failure, and his unyielding commitment to execution speed and quality. This book is a masterclass in getting things done, Musk-style.

Are you ready to take the plunge? To dive headfirst into a narrative that defies the norm, challenges the status quo, and offers an unprecedented glimpse into the life and mind of one of our era's most significant figures? Buckle up. This isn't just a story. It's a journey into the future, and Elon Musk is at the helm. Let's get started.

PART 1: DISRUPTION

1. PayPal's Revolution: The Birth of Disruption

AT THE TURN of the millennium, when the Internet was rapidly gaining traction, a young entrepreneur named Elon Musk eyed an opportunity in the realm of online financial transactions. At that time, many viewed online financial transactions as uncharted and dangerous territory. But Musk saw it differently: to him, it was an arena primed for disruption.

Musk, along with his co-founders, launched PayPal, a service that aimed to revolutionize the way people transferred money online. They faced a daunting task; it wasn't just about introducing or offering a new service—it was about changing deeply ingrained consumer behavior and confronting a well-established financial industry resistant to change.

PayPal had enormous obstacles from the start. Trust in online transactions was low, regulatory environments were intricate and hostile, and the technological infrastructure to support such a service was, at best, nascent.

Yet Musk did not waver. His strategy was grounded in one principle: to make online payments as simple and secure as possible. He knew that to disrupt, one had to think big, and big thoughts often require big risks.

The PayPal team leveraged the power of simplicity and accessibility combined with secure encryption technology to create an easy-to-use digital wallet. Musk understood that widespread adoption hinged on trust. According to a Wall Street Journal piece from 2002, Musk employed a referral strategy to create a network effect, giving both the referrer and the referee monetary incentives. This stroke of genius kick-started PayPal's exponential growth.

PayPal's disruption was not merely technical but also psychological. As Harvard Business Manager noted, it challenged traditional notions of money transfer, urging people to see the Internet as a legitimate space for financial transactions.

In the realm of disruption, PayPal was a game changer. It established the norm for online transactions and paved the way for the present financial tech revolution. As tech expert and New York Times contributor Kara Swisher aptly put it, "PayPal was not just a company; it was a statement against the old guard."

Musk was aware of PayPal's transformational potential. In an interview with The Economist, he observed, "We showed that if you can make a service convenient, people will change their habits."

As PayPal grew, it empowered small businesses, eased global transactions, and even catalyzed the gig economy. Its disruption paved the way for other innovative financial technologies, such as cryptocurrencies—a realm Musk himself would later venture into.

Reflecting on PayPal's journey, one can glean a deeper understanding of Elon Musk's approach to disruption: envision the impossible, embrace the challenges, and enact change relentlessly. Musk's journey with PayPal involved more than just creating a prosperous business; it also involved rewriting the rules and demonstrating that the existing quo could and should be contested.

In this light, tech investor Peter Thiel's reflection on Musk offers the most poignant summation: "Elon doesn't disrupt industries. He disrupts paradigms."

As we journey further into Musk's disruptive ventures, consider this: What are the paradigms in your own field or life that you've accepted as unchangeable? What if, like Musk with PayPal, you dared to challenge them? What could be your 'PayPal' moment?

Our journey into Elon Musk's world of disruption has just begun and PayPal is the first milestone. Hold on tight. As Musk once said, "When something is important enough, you do it even if the odds are not in your favor." Are you ready to defy the odds?

Having explored the inception and growth of PayPal, let's delve deeper into the ripple effect this revolutionary company has made in our society and economy.

Through PayPal, Musk demonstrated that major industries could be challenged and transformed. PayPal became a model for disruptive innovation, a beacon of what could be achieved when audacity meets technology. It sparked a wave of new ventures in FinTech, paving the way for platforms like Venmo, Zelle, and Stripe, thereby decentralizing financial power and making finance more accessible to everyone.

Furthermore,

PayPal's success was a tribute to Musk's risk-taking and resilience, which has become an integral component of his entrepreneurial philosophy. As he faced regulatory hurdles, skeptical consumers, and technological setbacks, he didn't back down. Instead, he pressed forward with a vision that would ultimately change the way the world does business.

The Harvard Business Manager highlighted Musk's uncanny ability to identify and exploit gaps in the market, stating, "Musk saw a problem, formed a solution, and created a new market."

This ability to innovate and disrupt wasn't just a result of keen business acumen; it was driven by a fervor to make people's lives easier and more efficient.

His 'never back down' attitude has not only been the cornerstone of his disruptive approach but also an inspiration to emerging entrepreneurs worldwide. His former PayPal colleague, Reid Hoffman, once remarked in a New York Times interview, "Elon's work at PayPal has shown us that no industry is impervious to disruption. His belief in his ideas, against all odds, is what sets him apart."

As we look back at PayPal's groundbreaking journey, we are reminded that disruption is never a smooth ride. It's an uphill battle against doubt, resistance, and uncertainty. Yet, the legacy of PayPal teaches us that the fruits of this struggle can shape industries, alter consumer behaviors, and catalyze economic transformations.

Through his unwavering commitment and strategic execution, Musk has proven that disruption is not merely a business strategy, but a mindset. A mindset that challenges the status quo, welcomes setbacks, and always strives to break new ground.

In wrapping up this chapter, let us reflect on Musk's own words during a 2018 interview with The Economist: "Some people don't like change, but you need to embrace change if the alternative is disaster." As we move forward, ask yourself, where are the stagnant waters in your life or your industry that need the ripple of disruption?

Our journey through Musk's disruptive ventures has only just begun. PayPal was the first step, the 'genesis of disruption,' if you will. There are many more chapters to this story, and as we navigate through them, remember Musk's approach to challenges—relentless, fearless, and visionary.

2. SpaceX: Shattering Space Monopolies

In 2002, Elon Musk turned his gaze skyward. Fresh off the seismic disruption of online financial transactions with PayPal, Musk launched SpaceX, a venture that would aim its rockets at a monopoly that had held the space industry captive: government agencies. With SpaceX, Musk sought to democratize space, bringing it within the reach of private entities and eventually, everyday individuals.

SpaceX was an audacious venture. The challenges were astronomical, literally, and metaphorically. Musk was aiming to disrupt an industry that had previously been dominated by powerful governmental entities such as NASA and Roscosmos. As tech expert Neil deGrasse Tyson noted in an article for The New York Times, "Elon Musk sought to take on a Goliath that had, until then, been challenged by no David."

The space industry was not just a monopoly—it was a technological fortress. It was an arena where failure had a huge cost. Furthermore, Musk was not merely planning to compete; he intended to slash costs dramatically, promising to make space travel as common as air travel. The world was skeptical, yet Musk remained unfazed. In the face of overwhelming odds, he stated,

"We're going to make it happen. As God is my bloody witness, I'm hell-bent on making it work."

The strategy was as revolutionary as the vision. SpaceX was committed to vertical integration, keeping the design and manufacturing of almost all components in-house, thereby driving down costs and ensuring quality control. Yet, what set SpaceX apart was its relentless pursuit of reusable rockets. According to The Wall Street Journal, Musk believed that the real key to opening up space was making rockets as reusable as airplanes, a concept that was revolutionary at the time.

The entire world was watching with bated breath as SpaceX had a spate of high-profile mishaps. Despite this, Musk and his team persisted, learning and iterating after each setback. As Musk famously said, "Failure is an option here. If things are not failing, you are not innovating enough."

In 2008, this grit bore fruit. SpaceX became the first privately funded company to send a payload (the Falcon 1) into orbit. The monopoly had been broken.

The success of SpaceX's has had a dramatic and disruptive influence on the space industry. It ushered in a new era of commercial space travel, prompting a renewed interest in space exploration worldwide. Companies like Blue Origin and Virgin Galactic soon followed suit, kicking off a new space race dominated by private sector.

SpaceX's disruption extended beyond space travel. It inspired a generation of entrepreneurs and rekindled our collective fascination with space. "SpaceX not only disrupted a monopoly," said Harvard Business Manager, "it reignited our dreams of the stars."

Musk's venture into the cosmos offers numerous insights. It demonstrates the need of a clear audacious vision, necessity for constant innovation, and influence of disruptive thinking. It proves that no sector is immune to upheaval, not even one as sophisticated and dangerous as space exploration.

Reflecting on SpaceX's journey, Musk offered a nugget of wisdom that encapsulates his disruptive ethos: "When something is important enough, you do it even if the odds are not in your favor."

Upon shattering the monopoly, Musk's SpaceX continued to innovate, proving to the world that his vision was not a one-time wonder, but a sustainable groundbreaking enterprise. The Falcon Heavy, Dragon spacecraft, and the awe-inspiring Starship showcased Musk's relentless pursuit of advancement. With each successful launch and each safe return of a reusable rocket, Musk was not just validating SpaceX's disruptive nature but also advancing humanity's understanding of space.

Perhaps the most transformative effect of SpaceX's disruption was its ability to inspire and drive change on a global scale. SpaceX's achievements created a wave of excitement around space exploration, compelling other nations to reassess their space programs. According to an article by The Economist, SpaceX's approach could be seen mirrored in other countries' space agendas. "From Europe to Asia, space programs are being revitalized and restructured, partly in response to SpaceX's advancements," the article stated.

However, the voyage can't be called voyage without its difficulties. There were failed launches, design challenges, and countless regulatory hurdles. The road to disrupting a monopolistic industry like space exploration was as rough as one might expect. But Musk's resilience and his commitment to his vision were unwavering. In an interview with the Wall Street Journal, Musk reiterated his faith in the power of tenacity and persistence, "If you get up enough times, you'll make it."

SpaceX's foray into space also offered an unprecedented commercial opportunity – satellite internet. The launch of the Starlink project aimed to provide high-speed internet across the globe, further illustrating Musk's ability to identify and capitalize on disruptive opportunities within an existing disruption.

SpaceX's legacy is ongoing and evolving. Yet, as we reflect on its journey thus far, a few key insights stand out. SpaceX is proof that no domain, no matter how complex or monopolized, is immune to disruption. It showcases the transformative power of a bold vision and the impact of tenacious execution. It underscores that disruption is not just about breaking down old paradigms but also about building new ones that propel us forward.

As we close this chapter, let's ponder over a thought shared by Musk during SpaceX's 10th anniversary, "I think it's possible for ordinary people to choose to be extraordinary." What SpaceX demonstrates is that this is not just a motivational quip but a testament to what is achievable when one dares to disrupt.

SpaceX's story encourages us to ask ourselves: What monopolies in our lives or in our industries do we accept as unchallengeable? And what could we achieve if we decided to be the disruptors instead of the disrupted?

As we navigate further into Musk's universe of disruption, remember this: Disruption begins with a spark of audacious vision, but it doesn't end there. It's fueled by tenacity, risk-taking, and an unyielding commitment to making that vision a reality.

3. Tesla: Accelerating the Electric Revolution

Elon Musk set his sights on another sector entrenched in convention and dominated by established players (automobile industry) in 2004, having already revolutionized the domains of online finance and space travel. Musk joined Tesla Motors as Chairman, later becoming CEO, with a singular mission: to accelerate the advent of sustainable transport by bringing compelling mass-market electric cars to the market.

Similar to SpaceX and PayPal, Tesla faced enormous challenges. The automobile industry was mature, saturated, and controlled by multinational giants with deep pockets. Moreover, the idea of electric vehicles (EVs) wasn't new, but previous attempts had resulted in cars that were lackluster, slow, and had limited range. As noted in The Wall Street Journal, the industry was skeptical if not outright dismissive of Musk's ambitions, considering them to be either wildly optimistic or profoundly naive.

Yet, Musk had a vision and a strategy. Tesla wouldn't just make electric cars; they would make the best cars that just happened to be electric. Tesla's Roadster, its first vehicle, wasn't a sedate, boxy, golf-cart. It was a sleek high-performance sports car that shattered preconceptions about what an electric car could be. Musk's strategy was not just to compete with traditional car manufacturers but to change the narrative around electric cars.

But breaking into the auto industry proved tougher than anticipated. According to a New York Times article, Tesla came incredibly close to bankruptcy in 2008. One of many dramatic events in Tesla's history, it was saved at the last minute by a fundraising round that ended on Christmas Eve. When asked about this period, Musk reflected, "The odds of a car company succeeding were incredibly low, but we did it anyway."

The advances made by Tesla extended beyond its automobiles. They altered the method of purchasing cars and selling directly to consumers rather than through dealers. They built a vast network of charging stations, addressing one of the biggest concerns around electric vehicles: range anxiety.

Slowly but surely, Tesla began to win over skeptics. The Model S, launched in 2012, was named the Car of the Year by Motor Trend. Tesla's market valuation surpassed conventional automakers by 2020, a clear indication of the disruptive impact it had on the industry.

The most transformative aspect of Tesla's disruption lies in its ripple effect across the auto industry and beyond. It spurred other automakers to fast-track their electric vehicle programs. It invigorated the renewable energy sector, propelling developments in battery technology and solar power. As stated in Harvard Business Manager, "Tesla's impact extends far beyond just the auto industry. It has driven a broad-based shift towards sustainability."

So, how did Musk steer Tesla through all the speed bumps and roadblocks? His philosophy was simple, yet powerful: "When you struggle with a problem, that's when you understand it. Once you understand it, you can solve it."

As we reach the end of this chapter, let's take a moment to reflect. If Elon Musk hadn't been willing to challenge the status quo, to risk failure in an industry where success seemed impossible, we wouldn't be witnessing the electric revolution on our roads today.

Musk's journey with Tesla serves as a vivid reminder that disruptive innovation isn't just about technological breakthroughs - it's about seeing the world not as it is, but as it could be. It's about having the courage to drive change even in the face of skepticism and adversity.

As we gear up for the next chapter in Elon Musk's journey, ask yourself this: What established beliefs and conventions are you willing to challenge? What's your vision for the 'impossible,' and what disruptive route will you drive to get there as the world started seeing the Tesla logo on roads? it became clear that Elon Musk had not only built a disruptive company, but also had spearheaded a global movement. Tesla's Supercharger network, a grid of high-speed charging stations, gave EV owners the freedom to drive long distances. It was another example of Musk's foresight in transforming infrastructures to support his disruptive ventures.

Then there was the Gigafactory, Tesla's battery production plant in Nevada. As reported by The Economist, it was seen as Musk's strategic move to control the supply chain and reduce battery costs. The Gigafactory symbolized Tesla's resolve to overcome challenges, drive efficiencies, and control its destiny.

With Tesla, Musk did more than just bring an innovative product to the market; he shifted societal attitudes towards sustainable energy. The increased awareness and demand for clean energy solutions that Tesla created were arguably among its most profound disruptions.

Tesla's journey also underscored Musk's belief in radical transparency and open-source philosophy. In a surprising move, he made Tesla's patents freely available. "We believe that Tesla, other companies making electric cars, and the world would all benefit from a common, rapidly-evolving technology platform," Musk shared in a blog post. This bold decision challenged traditional business strategies and further highlighting Musk's focus on accelerating sustainable energy over short-term competitive advantages.

But this journey was not journey without its roadblocks. Production delays, quality control issues, financial strains, and Musk's own combative public persona at times threatened to overshadow Tesla's accomplishments. However, they never detracted from the fact that Musk's audacious vision was turning into a reality.

In conclusion, Tesla's story reminds us that disruption is seldom a smooth ride. It's a winding road with steep climbs and sharp turns. It demands resilience, creativity, and the audacity to take risks. But as Tesla demonstrates, the destination is worth the journey.

Reflecting on Tesla's impact, Musk said in a 2020 interview, "We've certainly been through our fair share of difficulties, but the impact we've had is proof that daring innovation pays off."

With Tesla, Musk set out to do more than disrupt an industry; he wanted to help humanity transition to sustainable energy. He taught us that real disruption lies not just in creating a product, but in daring to redefine paradigms.

As we close this chapter, ask yourself: Are you willing to take the driver's seat and navigate the unpredictable roads of disruption? Are you ready to accelerate towards a sustainable future and transform the world in the process?

Indeed, when it comes to disruption, Elon Musk has demonstrated that the trip is just as important as the goal. As we continue on this interesting adventure, keep in mind that each disruptive step we take puts us one step closer to a better, more sustainable future. Buckle up, for the journey continues.

4. The Boring Company: Radical Urban Transformation

As we continue to traverse Elon Musk's journey of disruption, we find ourselves venturing into the literal and figurative depths of innovation with The Boring Company. Born from Musk's frustration with Los Angeles' infamous traffic congestion, The Boring Company represents a visionary answer to an everyday problem.

The Boring Company's mission is simple: to eliminate the bane of traffic by creating a vast network of underground tunnels, enabling swift, and efficient transportation of people and goods. The New York Times labeled it as "another high-stakes gamble from Musk," but as with all his ventures, Musk's audacity was only matched by his strategic execution.

The company's disruptive innovation is centered around a two-fold strategy. First, the development of Tunnel Boring Machines (TBMs) that are faster, more efficient, and less costly than conventional models. Second, creating an effective system for passenger transportation within these tunnels, something they've named "Loop."

Facing a slew of challenges, Musk needed to prove that tunneling could be cost-effective and practical on a large scale. He had to make it an appealing proposition to city planners, regulators, and the public. But if history has taught us anything, it's that Musk thrives on adversity.

In the face of skepticism, Musk pressed forward. "It shall be called 'The Boring Company'," he tweeted, using humor to allay doubts about the feasibility of his venture. As reported by the Wall Street Journal, Musk began tunneling operations in the SpaceX parking lot as a proof of concept, showcasing the feasibility of his vision.

The Boring Company's first commercial transportation Loop, completed in Las Vegas, serves as a testament to Musk's disruptive strategy. It demonstrated how rapid, convenient, and sustainable underground transport may become a daily reality. This was not just a disruption of transportation but also of the principles that govern urban planning and infrastructure development.

Musk sees a world with less traffic, less pollution, and less annoyance in the future. A world that is underground, where there is no weather and virtually infinite space. He is questioning the way cities are planned and how people get around by going below the surface and urging us to rethink how we move through and live in our cities.

As Musk once mused in an interview, "If you think of tunnels going 10, 20, 30 layers deep (or more), it is obvious that going 3D down will encompass the needs of any city's transport of arbitrary size." He encouraged us to think big, to look beyond the existing paradigms and visualize a different world.

The Boring Company is still in its early stages, but it stands as a symbol of Elon Musk's approach to disruption. It's a clear message that no problem is too mundane, no challenge too great, and no field immune to innovation if approached with courage, imagination, and strategic precision.

The Boring Company's tale motivates us to reevaluate how much we accept the way things are. It teaches us that a disruptive mentality views challenges as chances for creativity and sees constraints as invitations for change.

So, as we tunnel our way to the end of this chapter, let us ask ourselves: What 'boring' problems in our own lives can we approach with a disruptive mindset? How can we break through the concrete of conventional thinking and carve new pathways towards innovative solutions?

Remember this as we go further into the layers of Musk's vision in the next chapters: disruption isn't just about shaking up sectors; it's about bravely probing beyond the surface, challenging the conventional wisdom, and discovering new avenues for innovation. Continue exploring Musk's innovative projects as we move forward.

When we consider The Boring Company's influence, we see that Musk's ingenuity rests in both rethinking possibilities and modifying reality. He actively creates the future rather than just speculating about it. A truth that is clear from his approach with The Boring Company.

Traditional tunnelling endeavors were infamously costly and sluggish. Musk, on the other hand, relished this challenge and set out to maximize efficiency. According to The Economist, The Boring Company sought a tenfold decrease in tunnelling expenses. Although this was no easy task, Musk was used to setting lofty objectives.

His strategy centered on improving TBM design and functionality. The Boring Company's TBMs, as Musk explained in an interview with Joe Rogan, were designed to simultaneously dig and reinforce tunnels, reducing delays and increasing productivity. Furthermore, they were also equipped with electric motors, a deliberate choice to minimize environmental impact and align with his broader vision of a sustainable future.

Contrary to critics' skepticism, these innovative modifications were not mere fantasies. In an article in the Harvard Business Manager, the authors noted that Musk's "approach to tunnel boring encapsulates his knack for identifying efficiency gaps in existing technologies and relentlessly pursuing their improvement."

Indeed, The Boring Company was not only disrupting the transportation industry but was also spearheading a revolution in construction and engineering. This was a testament to Musk's multidimensional approach to disruption. He didn't merely operate within industries; he transformed them, breaking down barriers and redefining boundaries.

Still, Musk's ventures were not without their fair share of hurdles. Regulatory challenges, public skepticism, and technical complexities were constant companions on his journey. However, as with his previous ventures, these obstacles only fueled Musk's determination.

As we chart the course of Elon Musk's journey, we are reminded that true disruption is born from an amalgamation of vision, innovation, and tenacity. It's about daring to look at the world differently and having the audacity to bring that vision to life. From PayPal to SpaceX, Tesla, and The Boring Company, Musk's disruptive ventures serve as a testament to his unyielding spirit and ceaseless ingenuity.

No sector is too established for innovation, no issue is too complex for a fix, and no concept is too audacious for implementation, as demonstrated by The Boring Company, which personifies the spirit of disruption. As we wrap up this chapter, let's ponder upon a quote by Musk himself, "When something is important enough, you do it even if the odds are not in your favor." The Boring Company is an embodiment of this mantra, a monument to Musk's unwavering belief in the power of disruption.

5. SolarCity & Tesla Energy: Shaking the Energy Sector

In the spirit of disruption, Elon Musk set his sights on an industry as old as civilization itself, yet ripe for a revolution energy. With SolarCity and Tesla Energy, Musk envisioned an interconnected ecosystem of renewable power, echoing the sentiment of a quote attributed to him, "We have this handy fusion reactor in the sky called the sun. You don't have to do anything. It just works."

SolarCity, co-founded by Musk's cousins in 2006 with Musk as the chairman, aimed to provide affordable solar power to millions of homes across the United States. The company's business model was revolutionary, offering solar panel installation with no upfront cost, thereby tackling one of the primary barriers to solar adoption – initial expense. As reported by The New York Times, SolarCity grew rapidly, at one point becoming the largest provider of residential solar panels in the United States.

Still, SolarCity encountered several obstacles- the highly regulated energy industry and the ingrained utility corporations' interests faced significant obstacles. Despite these challenges, SolarCity persisted, demonstrating Musk's tenacity in upending firmly-established sectors.

In 2016, Tesla announced its intention to acquire SolarCity, a move met with skepticism by some. But Musk had a broader vision – a seamlessly integrated sustainable energy solution. This is where Tesla Energy came into play.

Tesla Energy, with its innovative power storage solution – the Powerwall, was a key component of Musk's vision for a sustainable energy ecosystem. As per an analysis in The Wall Street Journal, the Powerwall, designed to store solar power for residential use, was a critical step in addressing the intermittent nature of solar energy.

Integrating SolarCity's solar power generation with Tesla Energy's power storage created a loop of renewable energy production and consumption, thereby completing the picture of a sustainable home. Musk's grand plan didn't just stop at individual households. The Economist noted that Musk envisaged a world where entire cities could be powered with renewable energy.

Undoubtedly, the journey was fraught with challenges. Regulatory issues, technological hurdles, and the formidable task of altering consumer behavior were just a few of the trials Musk had to confront. But true to his spirit, Musk met these challenges head-on.

Looking beyond the story of SolarCity and Tesla Energy, we discern a deeper lesson about disruptive innovation. We may learn from Musk's strategy that disruption is about more than just inventing something new. It involves changing systems, perceiving things differently, and, most all, persistently working to make the impossibly possible..

So, how far would you go to disrupt a century-old industry? Would you dare to challenge the status quo and face down the giants? Elon Musk did. His vision for SolarCity and Tesla Energy reaffirms that in the face of disruption, no industry is too big, and no vision too audacious.

"The Boring Company" may sound like a play on words, but its mission is anything but dull. Its goal: to solve the problem of traffic congestion by creating a network of underground tunnels. It's an audacious vision, embodying Musk's philosophy of "thinking big," and one that is poised to disrupt city planning and transportation.

At a glance, the idea of burrowing a network of tunnels beneath a bustling metropolis seems an improbable solution. After all, tunneling technology has been around for centuries and has seen relatively little innovation. But, as The Economist noted, "If Elon Musk is involved, prepare for the improbable."

The difficulties faced by The Boring Company were enormous. Historically, tunnelling was expensive and took a long time. Red tape-filled bureaucracy stifled the process, but Musk, ever the shrewd businessman, exploited these obstacles as motivation rather than letting them deter him. He came up with an innovative way to do the actual tunnelling.

Musk proposed a design for faster, more efficient tunnel boring machines (TBMs) that could dig smaller, yet safer, tunnels. The New York Times reported that Musk's TBMs would not only bore the tunnels but also reinforce them simultaneously, eliminating time-consuming steps from traditional tunneling processes.

As ambitious as it is, could The Boring Company's approach truly disrupt urban transportation? Some critics were skeptical. Yet, as the Harvard Business Manager pointed out, skeptics have often underestimated Musk's ability to bring audacious visions to life. Indeed, test tunnels and plans for operational systems in places like Las Vegas and Los Angeles suggest that this venture is more than just a fanciful dream.

We conclude this chapter with a thought-provoking quote by Musk: "If you get up in the morning and think the future is going to be better, it is a bright day. Otherwise, it's not." SolarCity and Tesla Energy are a testament to Musk's belief in a brighter future. But what does your brighter future look like, and what would you be willing to disrupt to make it a reality? The answers may reveal the disruptor within you, waiting to break free.

As we move deeper into the world of Elon Musk, one thing is clear: his disruptive endeavors are not confined to any one industry. Now, let's turn our gaze to an enterprise that many considered as a whimsical fancy when it was first announced, yet today stands as a testament to Musk's relentless spirit of disruption - The Boring Company.

The Boring Company exemplifies Musk's drive to challenge the status quo. The idea of revolutionizing something as mundane and old-fashioned as tunneling might seem peculiar to some, but Musk teaches us that no area is immune to disruption and that transformative change often comes from the least expected places.

Elon Musk once said, "When something is important enough, you do it even if the odds are not in your favor." The journey of The Boring Company indeed mirrors this philosophy. Are you willing to take on something important even when the odds are stacked against you? If your answer is yes, you might just be on your way to becoming a disruptor in your own right. It's your move, dear reader.

6. SpaceX Starship: New Frontiers in Space Exploration

Elon Musk has always been a man who looks to the stars. His work with SpaceX has fundamentally altered the space industry, changing how we approach space travel and setting the stage for a new era of exploration and, possibly, colonization. The linchpin of this grand vision is a project as audacious as it is transformative: Starship.

Starship, the next-generation spacecraft being developed by SpaceX, is poised to disrupt the very notion of space exploration. Musk envisions it as a fully reusable vehicle that could carry up to 100 passengers to the Moon, Mars, and beyond. If successful, Starship will not just push the boundaries of what's possible; it will obliterate them.

The Wall Street Journal accurately encapsulated the ambition of this project, noting, "SpaceX's Starship represents the audacity of imagination coupled with the fearlessness of execution hallmarks of Musk's ventures." But, as with all disruptions, this audacious endeavor has not been without its challenges.

The project's technical complexities are immense. Starship's design involves a stainless-steel structure unlike anything seen in spacecraft before and its successful development requires innovations in heat shield technology, propulsion, and more. The challenges extend beyond engineering: securing funding, managing public expectations, and navigating regulatory frameworks are all part of the equation.

However, true to his nature, Musk has employed a unique strategy to tackle these challenges. He leverages the public's interest to gain support, routinely sharing updates on social media, and conducting presentations to draw attention to the project. He encourages a culture of rapid prototyping and iterative learning within SpaceX such as accepting failures as part of the path to success.

But what impact could this disruptive venture have? Beyond making space travel more accessible, the New York Times suggests that "the success of Starship could lead to a paradigm shift in our approach to inhabiting space." If colonization of other planets becomes viable, it may transform how we view our position in the universe, our approach to resource utilization, and our understanding of life itself. Elon Musk's vision doesn't stop at Mars. He envisions a future where humans become a multiplanetary species exploring the farthest reaches of the cosmos.

This chapter highlights the philosophical and ontological issues that Elon Musk's work poses in addition to the technological and commercial changes he is responsible for. His endeavors serve as a reminder that disruption involves challenging the current quo and broadening the boundaries of what is possible, not merely altering industries or technology.

As we close this chapter, consider this quote from Musk: "I think it is possible for ordinary people to choose to be extraordinary." Musk's work with SpaceX and Starship is extraordinary, indeed. As we look to the future and the potential for a new era of space exploration, we must ask ourselves: what extraordinary things will we choose to do?

Musk's audacious vision of the future invites us all to dream a little bigger, to not be afraid of failure, and to view challenges as stepping stones to success. What's the 'Starship' in your life?

Where will your audacious visions take you? Remember, the sky is no longer the limit.

A man who never shies away from a challenge, Elon Musk, has consistently confronted industries in desperate need of innovation. When he turned his gaze toward the energy sector, he didn't just aim to tweak the existing system; he sought to reimagine and revolutionize it. Enter SolarCity and Tesla Energy, two ventures that together aim to establish a complete sustainable energy ecosystem.

SolarCity, acquired by Tesla in 2016, focused on end-user solar power generation. Its unique business model allowed homeowners to install solar panels with no upfront costs, a breakthrough strategy that disrupted traditional utility business models and barriers to solar adoption. Harvard Business Manager described it as "the democratization of energy," a move that placed the power to generate renewable energy literally in the hands of consumers.

SolarCity faced myriad challenges. The initial skepticism towards solar energy, the convoluted landscape of energy regulations, and the capital-intensive nature of the industry were just some of the hurdles Musk had to overcome. Yet, he tackled these challenges with his trademark blend of audacity and strategic savvy. Through innovative financing models, consumer education, and the leveraging of government incentives for renewable energy, SolarCity began to change the narrative around solar power.

But Musk's vision for a sustainable energy future wasn't confined to power generation. With Tesla Energy, he embarked on a mission to address the equally vital aspect of energy storage. Tesla's Powerwall, a home battery system, and Powerpack, designed for commercial and utility-scale use, complement SolarCity's mission by offering efficient, reliable energy storage. Together, these ventures represent a potent disruption of the conventional energy grid.

So, what does this disruption mean for us? Firstly, it propels us closer to a sustainable energy future, something that's become increasingly vital in the face of climate change. Secondly, it's a testament to the power of integrated systemic thinking in driving innovation. By viewing power generation and storage as two sides of the same coin, Musk has challenged us to look beyond siloed solutions and think in terms of complete ecosystems.

But perhaps the most compelling aspect of this narrative is what it reveals about Elon Musk himself. At a time when most entrepreneurs were content to focus on a single industry, Musk took on multiple sectors simultaneously. His unflinching willingness to disrupt established industries and weather the inevitable storms of skepticism and resistance offer an invaluable lesson for aspiring entrepreneurs and innovators: Disruption isn't a one-time event, it's a mindset.

As this chapter draws to a close, it leaves us pondering a powerful question: In what ways can we embrace this mindset in our own lives and careers? Musk's journey inspires us to confront the status quo, to see potential where others see obstacles. His work with SolarCity and Tesla Energy beckons us to challenge established norms and to reimagine the future.

In the words of Musk himself: "When something is important enough, you do it even if the odds are not in your favor." As we navigate our own paths, we'd do well to keep this sentiment at the forefront of our minds, pushing us towards meaningful disruption, and ultimately, progress. So, ask yourself, what's the 'SolarCity' or 'Tesla Energy' in your life, and how will you disrupt it? Remember, the future is not set in stone; it's ours to shape.

7. OpenAI: Unleashing AI's Democratizing Power

When we think of artificial intelligence (AI), it often evokes images of sci-fi movies, autonomous robots, and futuristic technology. But for Elon Musk, AI was another realm ripe for disruption. His ambitious project, OpenAI, set out to ensure that artificial general intelligence (AGI)—highly autonomous systems that outperform humans at most economically valuable work—benefits all of humanity.

Elon Musk's disruptive approach to AI is rooted in his belief in democratizing access to technology. Rather than restrict AGI's potential to a select few, Musk has been vocal about the need to make AGI accessible to all. It's a daring idea, especially in an industry often characterized by proprietary rights and closed-source systems. The Harvard Business Manager aptly termed it as "open-sourcing the future."

Of course, this audacious endeavor was not without its challenges. A major concern was the potential misuse of AI technology. There were also questions around how to develop AGI in a safe ethical manner, especially given the absence of established norms or regulatory frameworks in this nascent field.

Musk's strategy, as with his other ventures, combined innovation with a steadfast commitment to his vision. OpenAI adopted a cooperative orientation, actively cooperating with other research and policy institutions to create a global community working together to address AGI's global challenges. This wasn't merely a strategy to mitigate risk—it was a call to collective action and a disruptive approach to a typically competition-driven industry.

The impact of OpenAI has been transformative. The New York Times described OpenAI's breakthrough language model, GPT-3, as a tool that "could democratize AI." By providing developers access to this advanced AI model, OpenAI is leveling the playing field, promoting innovation, and sparking a new wave of AI-driven applications.

Yet, beyond the technical implications, OpenAI's mission encapsulates a more profound disruption. It challenges the traditional closed-door approach to technological advancement. It disrupts not just how we build AI, but how we think about who should have access to it and who it should serve.

But what does this disruption mean for us, the everyday individuals? It signifies a shift towards a more inclusive future where the benefits of advanced technology are accessible to all, not just a privileged for a few. It's a reminder that disruption is as much about changing attitudes and norms as it is about revolutionizing industries.

As we reflect on OpenAI's journey and the disruption it represents, we might find ourselves asking: Where else could this approach apply? What other industries could benefit from this kind of open collaborative mindset? Musk's commitment to openness and cooperation in AI research encourages us to challenge traditional models of competition and secrecy.

OpenAI is a testament to Musk's disruptive thinking, a philosophy encapsulated in his words: "Some people don't like change, but you need to embrace change if the alternative is disaster." This echoes true for OpenAI's mission, urging us to disrupt before we are disrupted, especially in the face of a powerful technology like AGI.

Acting as an experienced ghostwriter for entrepreneur biographies, I shall now transition to the next chapter. This chapter delves into another of Elon Musk's ventures: Neuralink. Our focus will be on its audacious mission to develop ultra-high bandwidth brain-machine interfaces and the transformative effect it could have on humanity.

Titled "Neuralink: Pioneering the Path to Human-Machine Symbiosis," this chapter explores how Neuralink is pushing the boundaries of neuroscience, medicine, and technology. We'll examine the hurdles faced, the pioneering techniques employed, and the disruptive potential of such a venture.

Neuralink is the epitome of Musk's propensity for venturing into territories that many consider the realm of science fiction. It's his bet on a future where humans will be able to interface directly with machines, potentially transforming not just healthcare but how we live and interact with technology on a daily basis.

The journey of Neuralink has not been a walk in the park. From technical challenges of developing a safe, minimally invasive neural interface, to ethical considerations of integrating humans and AI, Neuralink has been threading uncharted territory. It has been tasked with tackling complex neuroscientific problems while ensuring compliance with stringent medical regulations.

But Musk has never been one to back down from a challenge. As with his other disruptive ventures, Musk employed an approach of bold innovation coupled with unyielding tenacity. The development of the Neuralink device, as reported by the Wall Street Journal, is a testament to this.

By focusing on a flexible, thread-like interface instead of traditional rigid implants, Neuralink is pushing the limits of current neurotechnology. The New York Times highlighted this significant departure from conventional methods as a move that "could revolutionize the field of neuroprosthetics."

Just imagine the profound implications of this technology. It could potentially help individuals with neurological disorders lead healthier and more fulfilling lives. It could transform the way we communicate, learn, and even experience entertainment. That's the promise of Neuralink - a future where humans are not just users of technology but an integral part of it.

However, Neuralink is more than just about human-machine interfaces. At its core, it's a powerful disruption of the notion of what it means to be human in an age of advanced technology. It takes risk of reimagining our connection with machines and poses provocative questions about our place in the evolving technological world.

So, what does this mean for us, as readers and potential disruptors? It's a call to think differently, to challenge existing boundaries, and to embrace the opportunities that come with disruptive innovation. It is a reminder of the transformative potential that lies at the intersection of different fields - neuroscience, AI, and biomedical engineering in Neuralink's case.

Neuralink, in its mission and ethos, serves as an embodiment of disruption - challenging the status quo, pushing technological frontiers, and daring to envision a future that most can't even fathom. As Musk once said, "You have to be pretty driven to make it happen. Otherwise, you will just make yourself miserable." As we move forward, may we find the drive within us to make our vision of a better future a reality.

8. Neuralink: Bridging AI and the Human Brain

Beneath the wide sky of his diverse, audacious ventures, one of Elon Musk's most fascinating pursuits stands unique, almost uncanny: Neuralink. In this chapter titled "Neuralink: Bridging Biology and Technology," we explore Musk's daring vision of merging the human brain with artificial intelligence, turning the realm of science fiction into reality.

Founded in 2016, Neuralink's ambitions are grand: to develop ultra-high bandwidth brain machine interfaces (BMI) that could seamlessly blend human cognition with artificial intelligence. According to Musk, this innovative interface could be the key to addressing several neurological disorders, augmenting human cognition, and even facilitating a symbiosis with AI.

However, the journey of this audacious venture is paved with considerable challenges. Neuralink is navigating the intricate confluence of neuroscience, biotechnology, and AI, where each discipline carries its own complexities. How can delicate neural networks interact safely with electronic interfaces? How can the vast data from the human brain be decoded into actionable insights?

Not daunted by these obstacles, Musk's strategy has been to attract top-notch talent from diverse fields, fostering an environment of radical innovation. His plan is to use machine intelligence, robotics, and material science developments to address the issue from many sides. This collaborative cross-disciplinary method has guided Neuralink to develop a robotic "sewing machine" that can insert thin threads into the brain with micron precision, a groundbreaking feat mentioned by the Wall Street Journal.

The transformative nature of this venture cannot be overstated. Imagine a world where paralysis could be overcome, where neurodegenerative diseases could be effectively managed, and where human cognition could be enhanced. This is the world Neuralink is trying to build disrupting the entire field of neuroscience and AI in its path.

But Neuralink is more than just a revolutionary tech company. It represents Musk's profound commitment to addressing existential risks associated with AI. Musk has consistently voiced his concerns about unchecked AI development, making headlines in the New York Times. With Neuralink, Musk aims to create a safeguard, a way for humans to 'keep up' with AI, thereby disrupting not only technology but also the discourse around ethical AI development.

What does this mean for us, the readers, and the aspiring disruptors? It signals the importance of fearless innovation in face of daunting challenges. Musk's venture prompts us to question the existing boundaries of technology and biology, pushing us to imagine what might be possible at their intersection.

Furthermore, Neuralink embodies the philosophy of 'First Principles Thinking,' a strategy often employed by Musk. Instead of accepting the constraints of existing approaches, Musk advocates for breaking down problems to their fundamental truths and building up solutions from there. This approach has helped Neuralink redefine what's possible in neurotechnology.

We learn more about Musk's overall goal of upending industries for the benefit of mankind as we investigate Neuralink's path. The idea of using AI to improve human cognition is more than just a technical achievement; it embodies Elon Musk's daring optimism about the possibilities of mankind. This conviction, together with his unwavering determination, is evidence of the book's main idea of disruption.

Neuralink, in its audacious vision and pioneering work, represents the spirit of disruption we seek to explore throughout this book. It challenges us to think big, to question, to innovate, and most importantly, to act. After all, as Musk himself shows, the future is a fascinating canvas, and we hold the brush.

The promise of Neuralink lies in a world where AI and humans coexist not just in harmony but in symbiosis. Neuralink seeks to redefine our relationship with AI, and in doing so, it redefines our perception of what is possible. Not only could Neuralink help to solve neurological conditions that have baffled medical science for centuries, but it could also provide a counter to the existential risk of uncontrolled AI. Through Neuralink, Musk, in his characteristic fashion, addresses a challenge by transforming it into an opportunity for humanity's progression.

Yet, the nature of Neuralink's venture into unprecedented territory brings forth important questions: how will society adapt to such advanced neurotechnology? How will this impact our notions of identity and privacy? What regulations should be in place for such a merger of human consciousness with AI? These thought-provoking questions underline the complexity of disruption. They remind us that disruption isn't just about the promise of progress; it is about acknowledging, addressing, and guiding the profound change it brings.

Additionally, as future entrepreneurs and leaders, the tale of Neuralink's disruptive innovation brings forth valuable lessons. It stresses the importance of cross-disciplinary innovation – how bringing together experts from various fields can lead to revolutionary breakthroughs. It also demonstrates how effective leadership can galvanize a team towards a seemingly insurmountable goal.

Reflecting on Neuralink's story, we are reminded of the audacity of dreams. For Musk, Neuralink isn't merely a company; it is a mission towards ensuring the future of human consciousness and potential. His vision serves as a powerful testament to his larger philosophy: thinking big, disrupting, and getting things done.

As we delve deeper into the life and ventures of Elon Musk, let us carry with us the essence of Neuralink's story. Let it remind us of the audacious potential within each one of us to dream, disrupt, and achieve. If the human mind can conceive it, can it not achieve it?

The visionary American architect, Buckminster Fuller once said, "You never change things by fighting the existing reality. To change something, build a new model that makes the existing model obsolete." In its quest to pioneer a new frontier in human cognition and AI, Neuralink seeks to do just that.

As we move to the next chapter, remember the tale of Neuralink, the daring venture at the cusp of technology and biology. Let it inspire you to question, dream, and innovate. And in this process, who knows, you might be the next disruptor laying the foundation for a future we have yet to imagine.

9. Hyperloop: Rethinking Global Transportation

ELON MUSK, the tireless architect of the future, is undeniably a figure of transformative ideas. The inception of Hyperloop, Musk's audacious plan to revolutionize the transportation industry, perfectly exemplifies this.

In 2013, Elon Musk proposed the Hyperloop, an ultra-high-speed ground transportation system. This system designed to transport people and goods in pressurized pods through a network of near-vacuum tubes at speeds that could surpass 700 mph and was not merely a departure from the norm. It was a paradigm shift that threatened to disrupt the existing transportation systems and setting the stage for a new era of mass transit.

However, Musk's path to the Hyperloop was not without its share of trials and tribulations. The project's audacious nature led to skepticism from experts and the public alike. Concerns about safety, infrastructure challenges, and regulatory hurdles were abundant. Yet, Musk's trademark tenacity shone through. He persisted channeling his knack for technological innovation and cultivated a team that shared his vision.

The strategy was simple and ingenious in equal measure: to develop a working prototype that would showcase the feasibility of the Hyperloop concept. By 2018, The Boring Company, another of Musk's ventures, began construction on a test tunnel in Los Angeles, thus bringing the theoretical Hyperloop a step closer to tangible reality. In addition to pursuing the technological frontiers, Musk also worked meticulously to address regulatory challenges, proving once again his ability to navigate complex business landscapes.

The disruptive potential of Hyperloop, however, extends beyond the realm of transportation. Consider the socio-economic impact. By slashing travel times and connecting distant locales as though they were neighboring districts, Hyperloop can fundamentally reshape how we live and work, offering a solution to urban overcrowding and potentially spurring economic growth in underdeveloped regions.

Reflecting on the grand vision of Hyperloop, Scott Galloway, a professor at NYU Stern School of Business, opines, "Musk isn't just redefining transportation; he's transforming our understanding of what is geographically possible." This statement embodies the core philosophy of Elon Musk – the spirit of disruption.

To give you a pilot's view, imagine being Musk, standing on the precipice of an idea so radical that it's unfathomable to many. The vision is crystal clear in your mind but transforming it into reality means battling a sea of skepticism and an army of challenges. The journey is arduous but the unwavering belief in the transformative power of your venture fuels the pursuit.

What drives Musk's relentless pursuit of disruption? The answer lies in his audacious belief in possibilities. As he famously stated, "When something is important enough, you do it even if the odds are not in your favor."

As we delve deeper into Musk's vision of disrupting conventional transportation, we are also compelled to ponder: how will our future look if we're bold enough to reimagine the present? How will our cities, our lives, and our world transform if we dare to embrace the audacious, the transformative, the disruptive? It's food for thought that begs for reflection and invites us to view our world through a broader lens.

As we close this chapter on a visionary leap into the future of mass transportation, we can't help but echo the words of the futurist Alvin Toffler, "The great growling engine of change - technology." Elon Musk, with his disruptive vision of Hyperloop, is undeniably at the helm, steering this great engine towards uncharted horizons.

In 2023, Hyperloop technology, despite its nascency, had begun to inspire and intrigue nations across the globe. Innovators, governments, and even rival transport corporations found themselves captivated by the possibilities that Hyperloop offered - attesting to the truly disruptive nature of Musk's vision.

Redefining speed and efficiency wasn't Musk's only goal with Hyperloop. He aimed to create a solution that was not only faster but also environmentally sustainable. In the era of escalating climate crises, this focus on clean renewable energy sources for powering the system was a visionary move. He envisioned Hyperloop pods that would be propelled by electric motors and solar power, thus minimizing carbon emissions.

Even amidst this groundbreaking innovation, Musk found himself grappling with challenges that were both technical and societal in nature. One of the biggest hurdles was convincing the public, policymakers, and potential investors about the viability and safety of Hyperloop. With the memory of high-profile failures in the transport industry skepticism was rife.

Addressing these concerns, Musk resorted to a strategy of openness and engagement. He shared the Hyperloop concept freely, inviting feedback, criticism, and innovation from the broader scientific community. This unorthodox move spurred a global conversation about the future of transportation and sparked a wave of innovation in the field.

The tech mogul's knack for dreaming big and executing grand plans in the face of adversity also inspired other entrepreneurs. Musk's relentless pursuit of his audacious vision despite significant challenges is a testament to his character and leadership style. In the words of Richard Branson, a fellow pioneer in transport innovation, "Elon is a risk-taker, willing to go for radical innovations that can make a significant difference."

As we close this chapter on a visionary leap into the future of mass transportation, we can't help but echo the words of the futurist Alvin Toffler, "The great growling engine of change - technology." Elon Musk, with his disruptive vision of Hyperloop, is undeniably at the helm, steering this great engine towards uncharted horizons. Like a seed planted in fertile soil, Musk's audacious idea has the potential to grow into a tree that can shade generations to come.

Looking back at Musk's Hyperloop vision, we find ourselves echoing the sentiment expressed by Wayne Gretzky, the hockey great: "You miss 100% of the shots you don't take." For Musk, Hyperloop is that shot — a shot at drastically transforming our transportation systems, a shot at leaving a sustainable world for our future generations, and ultimately, a shot at redefining what is possible.

10. Starlink: Connecting the Unconnected

Under the expansive umbrella of Elon Musk's technological pursuits, one particularly ambitious project stands out – Starlink. An initiative of SpaceX, Starlink aims to construct a satellite network capable of providing high-speed internet to every corner of the globe, thereby quite literally connecting the unconnected.

The concept behind Starlink is as captivating as it is disruptive. By launching a constellation of small, low Earth orbit (LEO) satellites, Starlink intends to offer broadband internet connectivity to underserved areas of the planet where access is non-existent or too expensive. Imagine the potential of such an endeavor. Access to reliable internet can transform lives, boost economies, enhance education, and bridge the digital divide.

However, as with any Musk venture, the path to realizing this disruptive vision was fraught with challenges. First was the formidable technical challenge of launching and maintaining thousands of satellites. Next came the regulatory hurdles – negotiating complex and varying international space laws was no small feat. Then, the environmental concerns related to space debris and light pollution needed addressing. Nevertheless, undeterred by the scale of these obstacles, Musk, as always, soldiered on.

Strategy in this complex scenario involved several key facets. Technically, SpaceX drew upon its established prowess in reusable rockets to make satellite launches affordable. Furthermore, the company adopted a phased deployment approach launching satellites in batches and gradually building up the constellation.

From a regulatory standpoint, SpaceX engaged in proactive dialogues with international space agencies and governments. To address environmental concerns, the company committed to making their satellites "dark-sky friendly" and incorporated an autonomous collision-avoidance system.

Starlink is not just disruptive; it's transformative. A fully functional Starlink system could drastically alter the telecommunications landscape and potentially ending the monopoly of large internet service providers. This could lead to more competitive pricing, higher speeds, and most importantly, democratization of internet access.

Imagine being in Musk's shoes at this juncture, steering a project that could be the panacea for global digital inequality. The magnitude of challenges may be overwhelming but so is the potential impact. It is this balance that seems to fuel Musk's disruptive journey.

Reflecting on his relentless push towards such audacious goals, Musk once said, "If something is important enough, even if the odds are against you, you should still do it." Starlink seems to be a manifestation of this philosophy - a significant step in the audacious march towards global connectivity.

The present is theirs; the future, for which I have really worked, is mine." The relevance of this statement to Musk's vision is uncanny. Starlink represents a profound shift in how we perceive internet connectivity, a future-focused venture that is quintessentially Musk.

As we delve deeper into this journey of disruption, one cannot help but wonder – how will universal internet access change the world? What opportunities will it unlock? How will it transform lives, economies, and societies? Pondering these questions, we begin to comprehend the colossal potential that Starlink holds and appreciate the audacity of Musk's disruptive vision.

This vision is driving Musk's relentless pursuit of the Starlink project. It might be worth asking ourselves: What does global internet connectivity mean to a small farmer in a rural community in Africa or to a schoolgirl in a remote village in Asia? The answer lies in the heart of Musk's mission - empowering the unconnected with information access, thereby enabling a world of opportunities.

Under Musk's leadership, SpaceX has been systematically overcoming the engineering and regulatory challenges that lay ahead. A crucial strategy Musk employed was to view these seemingly insurmountable challenges as engineering problems that are solvable with enough time and resources.

This pragmatic approach drew praise from several quarters. Renowned tech journalist Kara Swisher noted, "Musk doesn't merely innovate; he solves problems by entirely reframing the situation, taking on challenges others dismiss as impossible."

At this juncture, we must recognize the inevitable disruptions that Starlink could bring. The traditional telecommunications sector, for example, might find itself having to compete with a superior and more accessible service. Economically too, the impact would be substantial as connectivity becomes a facilitator of growth for previously isolated communities.

Starlink's transformative potential is thus vast and varied. The promise of affordable, high-speed internet to every corner of the world is poised to unlock unprecedented opportunities in education, healthcare, commerce, and social mobility. To put it into perspective, it is like handing over the keys of a vast library to someone who previously had access to only a handful of books. It's no less than a paradigm shift in information accessibility and digital inclusivity.

Ending this chapter, it is fitting to contemplate a statement from Musk himself: "When something is important enough, you do it even if the odds are not in your favor." Starlink is indeed important - not just to Musk, but to the vision of a connected and inclusive global community. It reiterates the power of 'thinking big' and how it can pave the way for disruptive transformations.

Musk's daring Starlink concept invites us to consider the limitless possibilities that universal connection may offer. If Starlink is successful, it might become a dazzling light of hope for billions of people, demonstrating that when technology and humans join together, the possibilities are unlimited.

We will continue this exciting journey of disruption in the following chapters, but for now, let's ponder - How could we leverage such connectivity in our respective fields? What would we do if we had the power to connect the unconnected? The future is open, and thanks to visionaries like Musk, it seems brighter than ever.

11. The Disruption Mindset: Musk's Blueprint

As we wind our way through Elon Musk's maze of disruptive companies, we see a similar thread: an unshakable philosophy of disruption that fuels creativity at every step. This last chapter connects the incredible adventures of Tesla, SpaceX, Hyperloop, and Starlink, illustrating how Musk's attitude can serve as a beacon for aspiring entrepreneurs and innovators.

Musk's business philosophy is summed up by his unshakable confidence in making the impossible feasible. Consider how he dealt with the numerous hurdles in his undertakings- whether it was the technological hurdles of developing reusable rockets for SpaceX or the cultural skepticism around the Hyperloop concept, Musk regarded them as issues to be solved rather than barriers.

In the face of adversity, his strategies have consistently been characterized by audacious risk-taking, bold innovation, and relentless execution. Drawing from his approach, young entrepreneurs can glean an important lesson - viewing challenges as opportunities for innovation can spur disruptive solutions.

Musk's ventures are not merely businesses; they are transformative forces reshaping entire industries. Tesla revolutionized the automobile sector, SpaceX is altering the dynamics of space exploration, Hyperloop is poised to disrupt mass transportation, and

Starlink seeks to redefine global connectivity. Through these ventures, Musk showcases the power of disruptive innovation in driving societal progress.

An essential aspect of Musk's philosophy, often overlooked, is his commitment to benefiting humanity. Every venture he embarks on aims to solve significant global issues – from sustainable energy with Tesla to global connectivity with Starlink. Aspiring innovators can learn from this - true disruption goes beyond creating successful businesses; it entails making a tangible impact on society.

His ability to continually challenge the status quo and drive transformative change has been admired by many industry leaders. Jeff Bezos, a fellow tech entrepreneur, once said, "Elon Musk is a guy who wants to go to Mars; that's ambitious. We need more of that in the world."

Reflecting on Musk's journey, a pertinent question arises: What is the essence of the Musk-style disruption? It is about envisioning a better future, embracing the risks involved, and forging a path towards that vision, however insurmountable the obstacles might seem.

It's a mindset that champions thinking big, going against the grain, and most importantly, getting things done. A mindset where disruption is the norm and transformation the goal.

In the words of Musk himself, "When something is important enough, you do it even if the odds are against you." This sentiment captures the spirit of Musk's disruptive mindset – a call to action that resonates with every entrepreneur and innovator seeking to make a difference.

As we conclude this part of the book, it is fitting to leave you with a thought: The world around us is a canvas of possibility, waiting to be disrupted, waiting to be improved. Inspired by Musk's journey, how will you paint your strokes of disruption on this canvas?

In the end, disruption is not just about transforming industries or creating groundbreaking products. It's about embracing a mindset, a mindset that questions, challenges, innovates, and dares to dream big. And in this journey of disruption, one finds not just the creation of extraordinary ventures, but also the crafting of a legacy that stands the test of time.

As we venture into Musk's world of disruption, the sheer scope of his vision leaves one awestruck. From revolutionizing the electric vehicle industry to democratizing space travel, reimagining mass transportation, and attempting to connect every corner of the globe. Musk has taught us that disruption is not about tweaking what already exists, but creating what should exist.

His ventures, while diverse, carry a unified theme - they disrupt, not for the sake of disruption, but to propel us towards a better, sustainable, and more connected future. Musk's approach to disruption is rooted in his unyielding belief in the potential of human innovation and his commitment to addressing some of humanity's most pressing challenges.

Consider SpaceX's audacious goal to make life multi-planetary. Musk not only disrupted the monopolized space industry but, in doing so, also opened up an entire new realm of possibilities. It's the kind of big picture thinking that allows him to see beyond the present, beyond the accepted norms, and envision a future where humans set foot on Mars.

This mindset, this vision-driven disruption is something every aspiring entrepreneur can take away from Musk's playbook. Being disruptive means daring to dream and doing everything in your power to realize that dream, no matter how impossible it might seem.

In an interview with '60 Minutes', Musk said, "Constantly think about how you could be doing things better and questioning yourself." This quote encapsulates the essence of his disruptive philosophy: a constant quest for improvement and a never-ending pursuit of innovation.

By now, we have witnessed the transformative power of Elon Musk's disruptive mindset. The question for every aspiring innovator and entrepreneur now becomes: How can you adapt and embody this mindset in your own journey?

To conclude this part of our exploration, let's reflect on a quote from renowned management expert Peter Drucker: "The best way to predict the future is to create it." Elon Musk's disruptive mindset embodies this philosophy, crafting futures that once only existed in the realm of science fiction.

As we look towards the future, let's keep Musk's audacious spirit of disruption in mind. What significant challenge would you like to address? How will you disrupt your field and contribute to shaping a better future?

In our investigation of Musk's disruptive path, one thing has become crystal clear: disruption is a mentality, not an idea. A attitude that fosters innovation, challenges the existing quo, and, most importantly, dares to imagine a better tomorrow. As we explore more into Elon Musk's incredible life in the future chapters, keep in mind that we all have the ability to disrupt, dream big, and get things done. After all, the future is a direction, not a destination, and the bold among us are its architects.

PART 2: THINKING BIG

12. SpaceX: Making Life Multiplanetary

Elon Musk, whose name is synonymous with boundary-pushing innovation, once professed, "You want to wake up in the morning and think the future is going to be great." This belief has been the driving force behind his audacious endeavors. It's in this spirit that Musk founded SpaceX in 2002 with an extraordinary goal: to make life multiplanetary.

The birth of SpaceX wasn't smooth sailing. The private space industry was practically non-existent, and the idea of a private entity building rockets seemed farfetched to most. Public opinion was skeptical, the aerospace industry was dominated by a few giants, and Musk himself had no formal education in rocket science. Nonetheless, Musk's indomitable will to dream and think big saw him through these challenges.

To understand Musk's strategy, one has to appreciate his unique approach to problem-solving: first principles thinking. As he explained in an interview with the New York Times, this concept involves boiling things down to the most fundamental truths, and then reasoning up from there. For SpaceX, this meant challenging conventional wisdom and redesigning rockets from scratch. His willingness to disrupt and take big risks was a crucial part of SpaceX's strategy.

This forward-thinking approach was evident in the creation of the Falcon 1, SpaceX's first rocket. While its initial three launches failed, the fourth successfully reached orbit in 2008. Rather than being deterred by these early failures, Musk and his team took them as lessons, iterating on their designs, refining their strategies, and pressing on with unyielding determination. This resilience underlined Musk's knack for getting things done, even in the face of overwhelming odds.

SpaceX's impact has been transformative, not just within the aerospace industry, but also in changing people's perception of what's possible. As former NASA deputy administrator Lori Garver commented in the Wall Street Journal, "SpaceX has demonstrated the success private industry can have in complex ventures." SpaceX's success disrupted an industry long dominated by governments and giant corporations, and heralded a new era of private space exploration.

Moreover, SpaceX's push towards reusable rockets has revolutionized the economics of space travel. The historic landing of the Falcon 9 in 2015 was a major breakthrough in this regard. Musk's vision of reducing space travel cost could make space more accessible and, eventually, enable humans to live on other planets.

Despite his exceptional accomplishments, Musk believes that the best is yet to come. His vision of colonizing Mars by 2050, as fantastical as it might seem now, illustrates his unwavering belief in thinking big. Musk once mentioned in an Economist interview, "History is going to bifurcate along two directions. One path is we stay on Earth forever and then there will be some eventual extinction event. The alternative is to become a spacefaring civilization and a multi-planet species."

In the grand narrative of Elon Musk's life, SpaceX stands as a testament to his audacity to dream, disrupt, and achieve the unthinkable. His journey encapsulates the idea that "Thinking Big" is more than just about having grand visions—it's about being resilient in the face of challenges, tenacious in pursuing goals, and unafraid of disrupting established norms to make those visions a reality.

As we look ahead, we find ourselves pondering the possibilities of an era where humans might become a multiplanetary species. If this prospect seems daunting or even impossible, remember that the SpaceX story started with one man daring to dream big. And isn't it the very nature of humanity to look beyond the horizon, to constantly strive for progress and exploration? As Elon Musk aptly put it, "When something is important enough, you do it even if the odds are not in your favor.

If this notion resonates with you, consider how you might apply Musk's brand of "thinking big" in your own pursuits. Might there be conventional wisdom in your field that warrants a fresh perspective? Are there disruptive strategies you could adopt to bring a bold vision to life?

Much like Musk's unwavering belief in SpaceX, confidence in your idea is crucial. Throughout SpaceX's journey, Musk faced multiple challenges that could have led to its downfall. From the three initial failed launches of Falcon 1, to the financial pressure of keeping the company afloat, to battling public skepticism, the odds seemed insurmountable. But Musk's conviction never wavered. His resilience and refusal to be discouraged by setbacks are hallmarks of his success and underscore his mindset of thinking big.

As we examine SpaceX's transformative impact, we can't ignore the wider implications of Musk's vision. By pushing the boundaries of space travel and reducing its cost, Musk is democratizing access to space. As SpaceX's reusable rockets continue to prove successful, we inch closer to the reality of becoming a multiplanetary species.

SpaceX's contributions extend beyond the realm of space travel. The company's pioneering work with satellite internet service, Starlink, aims to provide internet access to remote areas of the world, addressing a significant global inequality issue. Musk's ability to see how his ventures can intersect and drive progress in multiple domains is a testament to his big-picture thinking.

Turning our gaze towards the future, we are left to wonder about the potential ripple effects of Musk's vision. In a world where life becomes multiplanetary, how might this change our perspectives and priorities as a species? Will this lead to an unprecedented era of innovation and discovery?

As we draw to a close on this chapter, let us reflect on a quote by the famed science fiction author Arthur C. Clarke: "The only way to discover the limits of the possible is to go beyond them into the impossible." This encapsulates the spirit of Elon Musk's journey with SpaceX and his enduring capacity to dream, disrupt, and think big.

Allow this chapter to serve as a reminder of the value of dreaming big, challenging the current quo, and continuously pushing the limits of what's possible. Consider how you may dare to dream as big as Musk in your own endeavors. After all, we are, according to Musk, on the verge of history, poised to become a spacefaring and multiplanetary species.

13. Tesla: Driving into an Electric Future

An audacious journey that began in 2004 has since accelerated the world toward an era of sustainable transport. This remarkable adventure is none other than Elon Musk's visionary venture, Tesla. trip. In this chapter, we delve into Musk's unwavering dedication to realizing his vision of an electric future, which embodies the very best of big thinking.

When Musk first joined Tesla as chairman, the company was barely a blip on the automobile industry's radar. The obstacles were significant: traditional manufacturers were entrenched with significant resources, the public was skeptical, and electric cars were seen as impracticable. Despite these challenges, Musk was unwavering in his conviction that electric vehicles will be the way of the future.

Musk's strategy with Tesla was twofold. First, debunk the myth that electric cars were inferior. Second, create electric vehicles (EVs) so compelling that they would force the entire industry to take notice. As reported by the Wall Street Journal, Musk stated, "We had to get rid of the stigma of an electric car. We had to make it sexy, fun, and fast." The result was the Roadster, Tesla's first production vehicle and the first highway-legal EV to use lithium-ion battery cells.

But it wasn't enough for Musk to simply make an electric car. His ultimate vision was to bring about a large-scale transition to sustainable transportation. Thus, Tesla's mission: "to accelerate the advent of sustainable transport by bringing compelling mass-market electric cars to market." This vision led to the production of Model S, Model 3, Model X, and Model Y – vehicles designed to cater to a broad range of consumers.

The journey was not smooth, though. Musk was candid about Tesla's "production hell" during the ramp-up of Model 3. The company was on the brink of bankruptcy, a fact Musk later revealed in a New York Times interview. Nevertheless, his resilient spirit and relentless focus on problem-solving led to an impressive turnaround.

Tesla's impact has been nothing short of revolutionary. According to data reported by Harvard Business Manager, Tesla sold half a million vehicles in 2020 alone. Tesla's market value has also outpaced traditional auto giants, a testament to the transformative nature of Musk's venture. The Economist noted that Tesla's success compelled other automakers to expedite their electric vehicle development effectively disrupting the auto industry's status quo.

Yet the scope of Musk's thinking big extends beyond just automobiles. Tesla Energy, for instance, aims to facilitate renewable energy generation and storage. His vision of integrating renewable energy solutions with electric vehicles further exemplifies his unwavering commitment to a sustainable future.

As readers with entrepreneurial spirits and big dreams, consider Musk's approach. The challenges he faced were enormous, yet he never lost sight of his vision. What challenges in your own ventures might seem insurmountable today? And how can you, like Musk, stay resilient, think big, and continue to innovate for a better future? Just remember: every revolution begins with a spark, and Tesla is a shining testament to that truth.

As we reflect further on Musk's journey with Tesla, consider the transformational nature of his big thinking. It is one thing to invent a new product; it is entirely another to redefine an industry's status quo and influence the trajectory of the global economy. This is the magnitude of Musk's big thinking, a testament to his disruptive approach.

Perhaps the most profound aspect of Tesla's story is its wider societal implications. By championing electric vehicles, Musk has sparked a movement towards a more sustainable future—a shift that extends well beyond transportation. From promoting renewable energy to encouraging more sustainable consumer behavior, the ripple effects of Tesla's success are far-reaching.

By transforming the public's perception of electric vehicles, Musk has effectively driven demand for clean energy solutions. Consequently, we are witnessing increasing investments in renewable energy, battery technology, and related infrastructure. Tesla's disruptive impact is encouraging the evolution of entire industries and economies.

Yet, in typical Musk manner, he keeps having expansive thoughts. Tesla's autonomous driving technology now intends to change how we interact with automobiles. Additionally, Musk's plan to combine Tesla's energy solutions to build fully self-sufficient houses says a lot about his unwavering commitment to a sustainable future.

As you ponder Musk's extraordinary journey with Tesla, take a moment to reflect on your own entrepreneurial endeavors. What industry norms can you challenge? How can you encourage widespread change in your field?

It is important to note that Musk's journey had its share of setbacks and disappointments. But every obstacle was seen as a chance to learn, iterate, and advance—a crucial component of Musk's strategy for big-picture thinking. Musk once claimed in an interview with the Wall Street Journal that "Failure is an option here. If things are not failing, you are not innovating enough."

Let's close this chapter with a sense of eagerness and joy. One cannot help but ponder about the future prospects in a world that is adopting sustainable practices more and more. Elon Musk's path with Tesla may be summarized by Theodore Roosevelt's adage, "Do what you can, with what you have, where you are."

14. The Boring Company: Tunnels of Tomorrow

OUR WORST GRIEVANCES frequently serve as the seed for our wildest hopes. Elon Musk found the soul-crushing reality of Los Angeles traffic to be one such source of annoyance. Musk, a man renowned for his expansive ideas and determination to overcome obstacles, had a solution in his head. Introducing The Boring Company, his visionary initiative to transform urban mobility.

We can all agree: traffic is more than just a nuisance. It's a drain on our time, resources, and happiness. In 2016, the annual cost of traffic congestion in the US was $305 billion, a Wall Street Journal article reported. Musk, a thought-leader known for identifying problems and setting big goals, chose to tackle this issue with a characteristic twist: he would dig beneath it.

However, this involved more than just digging a few tunnels. Musk envisioned a sophisticated, extensive network of subterranean tubes that could move people, commodities, and vehicles at breakneck speeds. Of course, there were detractors and sceptics. People questioned its viability, safety, and overall chutzpah. But Musk remained unwavering. As IT expert John Naughton pointed out in a commentary for The Guardian, "Musk doesn't just think big; he thinks in dimensions most can't even fathom."

Such a great concept was difficult to put into action. Infrastructure projects are notoriously difficult, fraught with legal snags, complicated technical requirements, and exorbitant expenditures. The current technology was excessively pricey and sluggish. Unfazed, Musk adopted a two-pronged strategy: first, innovate to lower costs and expedite the tunnelling process; and second, simplify regulatory permits. In typical Musk flair, he intended to upset the established order.

What resulted was a company that developed a faster, more efficient tunnel boring machine, affectionately named 'Godot.' In a New York Times feature, Musk lauded the novel engineering design, which allowed continuous tunneling and building reinforcement simultaneously, vastly improving the efficiency of the process.

The first test tunnel in Hawthorne, California, was a testament to Musk's persistence and innovative prowess. Despite facing early skepticism, Musk showed the world that his 'pipe dream' was more concrete than abstract. With approval for an operational system in Las Vegas, Musk proved his disruptive vision could become reality.

The impact of The Boring Company goes beyond solving traffic congestion. It represents the transformative nature of Musk's ventures. Through SpaceX, Musk reimagined space travel. With Tesla, he challenged the fossil fuel industry. Now, with The Boring Company, he's reshaping our urban landscape. It's about more than convenience; it's about changing the way we perceive limitations.

This perspective defines Musk's approach to big thinking, as he once observed, "When something is important enough, you do it even if the odds are not in your favor." He is a doer as well as a dreamer, and The Boring Company is an example of this. It shows that if we have the will to get through the obstacles in our way, no desire is too big and no vision too outlandish.

Let's consider a thought as we learn more about Elon Musk's journey: What apparently unsolvable issues in our society may be solved if more of us dared to imagine as large as Musk? Keep in mind the man who saw a chance to explore the future when he looked at the congested streets while we debate this issue.

To quote esteemed economist Paul Romer, "A crisis is a terrible thing to waste." It invites us to take a page from Musk's book and turn our frustrations into the seeds of innovation. As we delve deeper into Musk's disruptive mindset in the coming chapters, let us seek inspiration and apply this big-thinking perspective to our own lives.

But how does Musk's vision translate into a world grappling with rapidly shifting realities and the urgent need for sustainable solutions? The Boring Company isn't just about transforming transportation; it's about building a future that's in harmony with our environment. With electric skates ferrying passengers and vehicles through the tunnels, the solution reduces surface congestion while simultaneously curbing CO_2 emissions. In an era where climate change has moved from a looming threat to a lived reality, Musk's big thinking is anchored firmly in sustainability, reflecting his multifaceted approach to problem-solving.

While disruption is Musk's modus operandi, there's also a keen focus on efficiency. Reflect on the Las Vegas Loop system developed by The Boring Company. A mere two years into its conception, the system was functional and serving thousands of passengers daily. In an industry where delays and cost overruns are the norm, Musk's ability to deliver is commendable. This raises the question: How can we inject such efficiency and focus into our projects and our lives?

His strategies hinge on embracing failure, encouraging innovation, and fostering relentless pursuit. "Failure is an option here. If things are not failing, you are not innovating enough," Musk famously said. This mindset underpins his ventures and is a strong factor behind his propensity to dream big and disrupt.

Each of Musk's ventures - Tesla, SpaceX, Neuralink, and The Boring Company - showcase the transformative power of ambition. They serve as a blueprint for all who aspire to create, innovate, and disrupt. And yet, they all connect back to a common thread: a refusal to accept the status quo and a passion to propel humanity forward.

It's worth considering how Musk's journey can impact our understanding of success and innovation. As we continue exploring his life, are there lessons we can internalize? Could Musk's fearless approach to dreaming big inspire a paradigm shift in our own lives?

Let's reflect on Musk's outrageous goal with The Boring Company as we get to the end of this chapter. In a world where practicality is frequently a limiting factor, Musk saw a time when vehicles would glide through tunnels, free from the restrictions of traffic. It is more than just a feat of engineering; it is a symbol of the influence of large thinking.

According to American science fiction author Robert A. Heinlein, "Everything is theoretically impossible, until it is done." Perhaps the true beauty of Musk's ambitious projects rests not just in their size, but also in their ability to motivate each of us to redefine our own definition of the impossibility.

15. SolarCity: Powering a Solar-Driven World

If you've ever glanced at a sunny sky and considered the power above you, you've shared a thought with Elon Musk. Known for his unconventional approach to addressing global problems, Musk's venture into renewable energy – SolarCity – was nothing short of visionary. SolarCity is not just about harnessing solar energy; it's about reimagining our future.

A plentiful, renewable source of energy is provided by the sun. However, solar energy has only lately become a significant participant in the world energy market. Here we have Musk's response to the energy crisis: SolarCity. His goal was to use clean, sustainable solar energy to power buildings, companies, and ultimately the entire planet.

SolarCity's journey was not without challenges. The company faced skepticism from those who doubted the viability of solar power as a mainstream energy source. As a Wall Street Journal article noted, "the solar industry, historically, has been a graveyard for investors." It required a leap of faith to believe in the potential of a sector fraught with financial risks and technological constraints.

Musk, though, appeared unfazed. His plan was to increase the affordability and accessibility of solar electricity. SolarCity removed the high upfront expenses of solar panel installation with a cutting-edge business strategy that provided solar leases and power purchase agreements. This strategy made solar energy affordable for typical homes, sparking a move towards renewable energy.

The impact of SolarCity's rise has been transformative, making solar power a viable and increasingly popular choice. According to a report by the U.S. Energy Information Administration, the solar industry has grown rapidly, with residential solar power increasing by more than 50% each year since 2010. A large part of this growth can be attributed to SolarCity and Musk's forward-thinking vision.

SolarCity's journey encapsulates Musk's propensity to think big. He looked beyond immediate hurdles and saw a future where energy consumption is sustainable, affordable, and in harmony with nature. To quote Harvard Business Review, "Musk isn't just building a business; he's trying to revolutionize an entire industry."

But changing ideas is just as important as upending industries. The initiatives of Elon Musk inspire us to challenge the existing quo and reinvent the world not as it is but as it may be. Musk himself made this clear when he said, "We're running the most dangerous experiment in history right now, which is to see how much carbon dioxide the atmosphere can handle before there is an environmental catastrophe."

It poses a challenge to each of us: Are we daring enough to think big, to envision a future radically different from our present, and to work relentlessly towards that vision? How can we, in our own spheres of influence, contribute to a more sustainable future?

As Musk's story unfolds, we see not only an entrepreneur creating businesses but also a visionary shaping the future. More than just a solar power firm, Musk's SolarCity catalyzed a global trend towards sustainable energy. His tale serves as a reminder that we have the ability to influence the course of history.

In the words of the great inventor, Thomas Edison, "I'd put my money on the sun and solar energy. What a source of power! I hope we don't have to wait until oil and coal run out before we tackle that." As we continue our exploration into the life of Elon Musk, let us take this sentiment to heart. How will we leverage the power of the sun, of renewable resources, to reshape our world? And more fundamentally, are we bold enough to envision a future powered by the immense potential that lies both above us and within us?

The impact of Musk's vision of a solar-powered world extends beyond the realm of technology and innovation. It's a vision that challenges our very way of life, inviting us to reconsider how we consume energy, interact with our environment, and plan for our collective future.

In a New York Times interview, Musk said, "We have this handy fusion reactor in the sky called the sun. You don't have to do anything, it just works. It shows up every day." This attitude reflects his knack for seeing simplicity in complexity, an approach that can inspire us all as we navigate our multifaceted lives.

As we have seen, Musk's journey with SolarCity wasn't always smooth sailing. He faced criticism, regulatory hurdles, and the initial hesitance of consumers to adopt solar power. Yet, he held firm, demonstrating a resilience and commitment that's worthy of admiration. So, how can we apply this same resilience in overcoming our personal and professional challenges?

It's also worth noting how Musk, a self-proclaimed "nano-manager," applies his leadership skills to his bold visions. As former SolarCity CEO Lyndon Rive noted in an interview with Harvard Business Manager, Musk's "intense, detail-oriented approach" was instrumental in overcoming the various challenges they faced. How can we adopt such an approach to ensure the successful execution of our projects?

The story of SolarCity also highlights Musk's ability to think globally. From SpaceX's interplanetary ambitions to Tesla's worldwide impact, Musk's dreams consistently transcend borders. As entrepreneur and LinkedIn co-founder Reid Hoffman once said, "Entrepreneurs like Elon Musk are transforming the globe. It's a wake-up call for us to be bold in our thinking."

Yet, Musk's grand vision isn't about achieving personal glory. It's about humanity's collective journey towards a sustainable future. It's about understanding that the decisions we make today will shape the world of tomorrow. This thought prompts an important question for all of us: What steps are we talking to contribute to a sustainable and equitable future?

As we conclude this chapter, we reflect on the journey of SolarCity – a testament to Musk's ability to dream big and disrupt the norm. Beyond the story of a company, it's a tale of audacious ambition, resilience, and an unwavering belief in a better, sun-powered future.

As we ponder on Musk's contribution to the renewable energy sector, a quote by futurist and philosopher Buckminster Fuller comes to mind: "You never change things by fighting the existing reality. To change something, build a new model that makes the existing model obsolete." And that's precisely what Musk did with SolarCity.

Looking ahead, let's carry this perspective with us: What existing models can we challenge in our lives? How can we, inspired by Musk's example, become not just consumers, but also builders and innovators of a brighter future? As we flip to the next page of Musk's life, let us open our minds to the boundless possibilities that big thinking can unleash.

16. SpaceX Starship - Aspiring for Galactic Colonization

"Life needs to be more than just solving problems every day. You need to wake up and be excited about the future, and be inspired, and want to live," Elon Musk once proclaimed. This credo reflects in one of Musk's most ambitious ventures: SpaceX, with its Starship rocket poised to transport humans beyond our home planet, to Mars.

The vision Musk has for SpaceX is as audacious as it is compelling: Establish a self-sustaining colony on Mars. It's a goal that redefines the term 'long-term planning.' The question is, why Mars? In Musk's view, becoming a space-faring civilization is not just an aspiration, but a necessity. He sees it as a safeguard against potential existential threats to human survival.

To accomplish this, Musk faced challenges on every front. Building a cost-effective, reusable rocket, obtaining governmental approvals, and more broadly trying to do something that no private company has ever done before. A 2022 Wall Street Journal report highlighted SpaceX's battles with the Federal Aviation Administration over its launch licenses. Despite these obstacles, Musk remained committed refusing to see failure as an option.

His plan for SpaceX was straightforward yet ground-breaking. By creating reusable rockets that could return to Earth after delivering their cargo, he sought to lower the often exorbitant expenses associated with space flight. Musk hinted at his far broader plans for SpaceX in a 2020 interview with The Economist, saying, "I think it's going to open up the solar system to humanity."

Through SpaceX, Musk has indeed revolutionized the aerospace industry, rekindling the public's interest in space exploration and making space more accessible. According to Harvard Business Manager, SpaceX's accomplishments have "disrupted a decades-old industry, inspired new startups, and attracted private capital to the space sector."

Yet, the impact of SpaceX and the Starship goes beyond the realm of technology and business. It challenges our perception of what's possible, igniting a collective imagination about the future. What might life look like on another planet? How will this transform our understanding of ourselves and our place in the universe?

As SpaceX continues to push boundaries, Musk's audacious vision serves as an emblem of a different way of thinking. A mindset that refuses to accept limitations, that dares to venture into the unknown. Reflecting on SpaceX's mission, renowned physicist Stephen Hawking said, "Spreading out into space... will completely change the future of humanity."

Musk's bold dreams for SpaceX beg several important questions. How do we as individuals or organizations maintain the courage and audacity to dream big even when facing seemingly insurmountable challenges? And how far are we willing to go to realize these dreams?

SpaceX is not merely a rocket company. It's a manifestation of Musk's profound optimism for the future, a testament to his ability to dream big and get things done. As we gaze into the stars, Musk encourages us to see not just the vast expanse of the cosmos, but the boundless possibilities of human potential.

SpaceX's successful launch and landing of reusable rockets represent not just a triumph of engineering but a victory for audacious thinking. Yet, despite these significant achievements, Musk's mind is perpetually fixed on the horizon specifically the reddish hues of Mars' landscape.

In Musk's vision, the Starship won't just ferry people to and from the International Space Station. Instead, it represents the cornerstone of his plan for making life multiplanetary. To this end, SpaceX's engineers are in an ongoing process of enhancing the Starship's design, incorporating lessons from each launch, and making iterative improvements. This demonstrates Musk's belief in learning through failure, a philosophy that can equally apply to our personal and professional lives.

However, one cannot overlook the considerable risks and criticisms associated with this ambitious project. "The first crewed mission to Mars could be a one-way trip," Musk said in an interview with The New York Times, underscoring the enormous risks involved. Critics argue that the focus should be on Earth and its mounting environmental issues rather than colonizing another planet. Musk is steadfast in his belief that mankind must evolve into a multiplanetary species in order to survive and advance.

Despite the hurdles, the potential payoff of Musk's vision is transformative. "SpaceX is the biggest game-changer for the future of humanity," opined renowned tech expert and futurist Ben Hammersley, "It has the potential to make us an interplanetary species." Moreover, the technological advancements made by SpaceX could have terrestrial benefits, from advances in materials science to enhancing Earth's satellite infrastructure.

The aspirational narrative of SpaceX reflects the compelling nature of Musk's leadership. His vision for a Mars colony acts as a powerful rallying call, inspiring his team, sparking public interest, and attracting investors. This highlights the importance of a compelling vision in achieving ambitious goals.

For Musk, going to Mars is not some irrational drive for planetary dominance. Instead, it's a crucial stage in the survival and evolution of humans. It forces us to reevaluate our position in the universe as we transition from a planet-specific species to a multi-planetary one.. It's a narrative that forces us to question: What would it mean for humanity to become a multiplanetary species? And what role do we want to play in this grand journey?

In concluding this chapter, let's ponder on a powerful statement by late astronomer Carl Sagan: "Imagination will often carry us to worlds that never were. But without it, we go nowhere." Musk's vision for SpaceX and Mars colonization embodies this spirit of imagination and exploration. As we continue this journey through Musk's life, let's keep this sense of audacity and wonder at the forefront. What world-changing ideas can we dream of? How will we strive to turn these dreams into reality, boldly venturing where none have gone before? With that, we look forward to exploring more facets of Elon Musk's extraordinary life in the chapters to come.

17. OpenAI - Cultivating AI for All

How does one strive to democratize the power of an entity as transformative as artificial general intelligence (AGI)? This question has found an audacious proponent in Elon Musk. Co-founding OpenAI, an organization committed to ensuring AGI's benefits for all. Elon dared to envisage a future where technology and humanity coalesce for mutual enhancement.

OpenAI's journey was laden with challenges. In a world where tech giants keep their AI advancements shrouded in proprietary secrecy, Musk's vision of offering AGI's benefits to everyone faced skepticism. Some feared that misappropriation of this powerful technology could lead to catastrophic consequences. Yet, Elon saw past these concerns. He envisaged a future where technology becomes an enabler and not a gatekeeper.

Elon's strategem for OpenAI pivoted around the principle of transparency. He believed that for the democratization of AGI to be meaningful, its development must not remain in the clutches of a select few. OpenAI pledged to provide public goods and share most of its AI research, a practice unheard of within the traditional tech landscape. How audacious it was to place the fruits of this cutting-edge research within everyone's reach!

Yet, Elon's gambit paid off. OpenAI's transparency ushered an era of unprecedented collaboration in the AI community. Researchers across the globe began contributing, further pushing the boundaries of what AGI could achieve. Consequently, OpenAI's impact has been profound, driving AI breakthroughs and increasing global access to these technological marvels.

The transformative nature of Elon Musk's ventures is evident in OpenAI's work. The organization's high-profile projects, such as the AI-powered language model GPT-3 have disrupted the tech industry's norms. Yet, Musk's influence extends beyond disruption; it lies in how he has reshaped the narrative surrounding AGI's potential. As Rana Foroohar, a renowned journalist for the Financial Times, aptly stated, "Musk's OpenAI is not just a technological venture; it's a radical reimagining of technology as a force for shared good."

How does Musk's broad thinking for societal benefit reflect in OpenAI's ethos? This takes us back to Elon's conviction that AGI can be a tool for collective progress. The stipulation that any influence over AGI's deployment will be used for humanity's benefit echoes this conviction. By championing broad distribution of benefits, OpenAI seeks to mitigate societal power imbalances stemming from AGI's uneven use.

The Muskian mantra of "thinking big" is embodied in the audacity of OpenAI's mission. It signifies a leap of faith into the uncharted territories of AGI's potential while upholding the banner of collective good. Yet, it is equally a reflection of Musk's unwavering conviction that humanity, when armed with the right tools, can surmount its greatest challenges.

Elon Musk's vision for OpenAI invites us all to question the status quo, to dream of a more equitable technological landscape, and to dare to shape that dream into reality. But what does this mean for you? How can you, as an entrepreneur, a tech enthusiast, or a curious reader, leverage the power of AGI for societal good?

As we delve further into the intricacies of Elon's thinking, we uncover the underlying thread that connects his ventures - a relentless pursuit of transformative ideas. Just as OpenAI has revolutionized the AI landscape, Musk continually seeks new avenues to enact meaningful change. This trait and unwavering dedication to progress is an embodiment of the spirit of innovation we all can imbibe.

"Most people overestimate what they can do in one year and underestimate what they can do in ten years," remarked Bill Gates, an insightful sentiment that Musk seems to have taken to heart. OpenAI is not merely a testament to the power of artificial intelligence; it's a bold commitment to the long-term potential of technology to augment our collective abilities and improve our world.

Musk's bold aim has challenged conventional wisdom by demonstrating that artificial intelligence is a practical tool for change rather than a sci-fi fantasy. His unrelenting pursuit of this vast idea serves as a role model for aspirant businesspeople and inventors all around the world.

The inception of OpenAI and its disruptive vision invites us all to take a leap, to embrace challenges, to learn, and to grow. It challenges us to view setbacks not as definitive roadblocks but as stepping stones towards a broader and more inclusive vision. This ethos forms the foundation of Musk's disruptive and transformative approach to technology and innovation. But what can we learn from it? How can we apply it in our own lives and industries?

For one, we can start by adopting a more expansive perspective. Just as Musk dared to imagine a world where AGI is accessible to all, we can start envisioning how our work can serve a broader purpose. We can challenge the status quo, push boundaries, and drive change in our own capacities.

But the story doesn't end here. The spirit of OpenAI is about more than just "thinking big" – it's about "acting big" too. It's about taking concrete steps to actualize our visions, just as Musk has done time and time again.

As we conclude this chapter, we invite you to ponder: How will you harness the power of disruptive thinking in your journey? How will you ensure that your work benefits a wider community, fostering a more inclusive and equitable future?

In the words of Elon Musk himself, "When something is important enough, you do it even if the odds are not in your favor." Let this be the clarion call to all aspiring changemakers. Dare to dream, dare to disrupt, and above all, dare to make a difference.

The story of OpenAI is not just about a technological venture; it's a story about the audacity of hope and the transformative power of big thinking. As we move forward, we carry with us the spirit of Musk's vision – a testament to the indomitable power of human creativity and the boundless possibilities that lie within our collective reach. The road ahead may be uncharted, but as Musk has shown us, with audacity and conviction, we can pave the way towards a brighter future.

18. Neuralink - Foreseeing a Symbiotic Future

In Elon Musk's world of "Thinking Big", no idea is too audacious, no vision is too grand. His venture to Neuralink, a neurotechnology company, is a testament to this philosophy. Envisaging a symbiosis between the human brain and artificial intelligence (AI). Musk seeks to enhance human cognition, creating a future were biology and technology merge seamlessly.

Musk founded Neuralink in 2016, with a vision that could easily belong in a science fiction novel. He wanted to enable humans to keep pace with AI's rapid advancements by integrating technology into our brains. By developing ultra-high bandwidth brain-machine interfaces, Musk envisions a future where humans can communicate telepathically, upload memories, or even download skills, much like in the movie 'The Matrix'.

The difficulties were numerous. The challenges looked daunting, from designing a device tiny enough to be implanted in the brain without invasive surgery to creating a user interface that could decipher complicated neurological data. However, as we've repeatedly seen, Musk is no stranger to overcoming obstacles.

The strategy was clear yet audacious: to design and build a scalable, implantable device that could read and write to the brain. As Dr. Tim Denison, a professor of Biomedical Engineering at Oxford

University, noted in a New York Times article, "Musk's ambition is to alter the trajectory of the human race."

Neuralink's impact is already starting to be felt, albeit in the early stages. The successful trials on animals and the recent FDA approval for human trials indicate a promising start. If successful, Musk's venture could revolutionize not only how we interact with technology but also how we understand and treat neurological conditions.

When we contemplate the wider ramifications of Neuralink's aim, the revolutionary nature of Musk's endeavors becomes even more obvious. The possible advantages include helping paraplegics regain movement and managing mental health issues. The ultimate objective is still to improve mankind and maintain it on par with artificial intelligence, which is daring and unabashedly Muskian.

But what does this mean for the future? Can we envisage a world where such a seamless integration of technology and biology is not only possible but beneficial? The answer, much like Musk's vision, is expansive.

As an entrepreneur or tech enthusiast, Neuralink's story encourages you to push the boundaries of what is possible. It challenges you to envision a future that might seem far-fetched today, but could become a reality tomorrow. It teaches you to embrace the unknown, to tackle daunting challenges head-on, and to persist in the face of skepticism.

"Thinking Big", in Musk's terms, is about imagining futures that others deem impossible and having the audacity to turn those visions into reality. It's about transforming our perception of technology from a tool to a partner and from an external entity to an integral part of our identity. It's about pushing the boundaries of human potential and continually raising the bar for what we can achieve.

In the words of Musk himself, "If something is important enough, even if the odds are against you, you should still do it." As we chart the unexplored territory of AI and human symbiosis, Musk's audacious vision serves as an inspiration to push the envelope of possibility.

In Neuralink, we find a manifestation of Musk's propensity to dream and think big. The venture, while still in its nascent stages, holds the potential to redefine our interaction with technology, blending the boundaries between biology and AI. But Musk's vision goes beyond simple technological integration. He envisions a future where technology enhances the human experience, augmenting our capabilities, and empowering us to keep pace with AI.

This audacious thinking, rooted in the bold integration of technology and biology, propels us into a realm where the possibilities seem boundless. Neuralink is not just about a product or a device; it's about crafting a future where technology and humanity coalesce, enabling us to transcend our biological limitations.

As Musk once stated in a Wall Street Journal interview, "The future is vastly more exciting and interesting if we're a space-faring civilization and a multiplanetary species than if we're not." This quote underpins Musk's overall philosophy. It's not merely about survival; it's about striving, growing, and constantly pushing the boundaries of what's possible.

With Neuralink, Musk is not just challenging the way we perceive and interact with technology; he's also challenging us as individuals and as a society. He's compelling us to think about our role in this technologically advanced world and the ways in which we can contribute to its growth.

As we conclude this chapter, a quote from the famous physicist Richard Feynman springs to mind: "What I cannot create, I do not understand." Musk's vision for Neuralink is an embodiment of this sentiment. By striving to understand and replicate the human brain's complex mechanisms, he seeks to create a future that holds untold potential for human progress.

We should remember that 'Thinking Big', as demonstrated by Elon Musk, is not simply about having grand ideas; it's about having the courage, the determination, and the tenacity to turn those ideas into reality. It's about challenging the status quo, pushing the boundaries, and refusing to be constrained by what's been done before. In the spirit of 'Thinking Big', we leave you with a thought-provoking question: In a world of boundless possibilities, where biology and technology can merge to create something extraordinary.

19. Hyperloop - Pioneering High-Speed Travel

ELON MUSK'S idea for the Hyperloop was born out of a fundamental dissatisfaction with the way things were. While many admired the advancements in digital technology, Musk focused on the real environment and how people move through it. Therein lay the germ for the Hyperloop because he envisioned the future of transportation not as gradual improvements but as revolutionary leaps.

Introduced in 2013, the Hyperloop was Elon Musk's bold solution to modern transportation woes. A high-speed transit system capable of travelling near the speed of sound, the Hyperloop proposed shuttling passengers and freight in pods through a vacuum-sealed tube, reducing the Los Angeles to San Francisco commute from six hours to a mere 35 minutes.

His vision, however, was met with an onslaught of criticism. Experts questioned the feasibility of the technology, the astronomical costs associated with building infrastructure, and the potential safety risks. Musk's answer was a testament to his fortitude. He released a 57-page white paper, inviting everyone, competitors and critics alike, to join him in solving these challenges.

Musk's strategy in the face of adversity was transparent collaboration. He encouraged open-source development, inviting other companies and research institutions to refine and realize his idea. It was an unconventional approach, but one that reflected Musk's belief in the importance of collective genius. His determination was infectious. Companies like Virgin Hyperloop and universities worldwide embraced the challenge, propelling the technology forward through research and development.

The implications of the Hyperloop are transformative, not just in terms of speed, but also in sustainability. By harnessing solar power and regenerative braking, the system could operate with a near-zero carbon footprint, making it a potential game-changer in the fight against climate change. In a world where time is money, the Hyperloop's ability to connect major cities within minutes could usher in a new era of globalization, revolutionizing the way we work, live and interact.

Musk's vision is vast and daunting, but it is not unfounded. We've seen him prove naysayers wrong before, like when SpaceX successfully launched and landed reusable rockets, a feat deemed impossible by many. His big thinking continues to inspire a generation of innovators and disruptors, motivating them to dream beyond what's immediately achievable and redefine the boundaries of possibility.

The Hyperloop is more than just a mode of transportation. It serves as evidence of Musk's daring and unwavering attitude. It serves as a reminder that in order to accomplish real progress, we must challenge the status quo, challenge conventional knowledge, and accept the challenges that come with initiating change.

As Musk himself stated, "When something is important enough, you do it even if the odds are not in your favor." His tenacious commitment to the Hyperloop epitomizes this ethos, encouraging us all to dream bigger, push harder and refuse to settle for mediocrity.

"Most people overestimate what they can do in one year and underestimate what they can do in ten years." - Bill Gates. A testament to the fact that with enough grit and unwavering resolve, a person can indeed disrupt the world, just as Musk has and continues to do. As we delve deeper into Musk's entrepreneurial journey, we shall examine more of these moonshot ideas, each one representing a challenge to entrenched systems and each one bearing the stamp of his commitment to transformative, sustainable innovation.

The Hyperloop is not just a fantasy. It is a physical representation of Musk's unwavering faith in a better future, his reckless sense of adventure, and his unwillingness to take "impossible" as an answer. Even while the project is still in the early phases of research, we must keep in mind that reusable rockets and electric automobiles, which Musk's efforts have advanced to the verge of becoming standard, were once futuristic notions.

His daring vision teaches us that progress isn't born from safety, but from a courageous willingness to venture into the unknown. Musk doesn't merely solve problems; he redefines them, turning insurmountable barriers into opportunities for innovation. In the face of doubters, his strategies remain consistent: harness the power of collective intelligence, be transparent about challenges, and remain unwavering in the pursuit of breakthrough solutions.

This perseverance has far-reaching effects, motivating future generations of businesspeople, inventors, and social activists to take on humanity's toughest problems head-on. The Hyperloop is proof of the power of big ambitions and the amazing things people can do when they dare to dream larger.

But how does one maintain the drive, the unyielding determination in the face of such mammoth tasks? What fuels the fire of innovation in a mind like Musk's? As we continue to explore his life and work, we shall delve into the very core of his relentless drive, a factor as integral to his success as his ability to think big.

In the words of the futurist and philosopher, Buckminster Fuller, "You never change things by fighting the existing reality. To change something, build a new model that makes the existing model obsolete." Musk's journey embodies this philosophy, as he continually disrupts industries not by mere incremental adjustments but by complete, groundbreaking overhauls.

His story serves as a clarion call, challenging each of us to embrace the audacity of big thinking. It's a reminder that the future is shaped by those willing to envision a radically better world and who, in the face of resistance, stand unwavering in their resolve to bring such visions to life.

As we journey further into Musk's remarkable narrative, we ask you, the reader, to reflect on your own potential for big thinking. What audacious ideas lie within you, waiting to be unleashed? How will you contribute to reshaping our world, just as Musk is doing?

Keep in mind that the goal is not to slightly enhance the performance of current systems. It's about coming up with brand-new ideas that supplant the previous ones. So let's consider this as we learn more about Elon Musk's life. Think large because it is where the ability to transform, innovate, and disrupt rests. For as Musk demonstrates, the impossible can indeed become possible when you dare to dream big.

20. Starlink - Global Internet Coverage

IMAGINE a scenario in which high-speed internet is available in every location on Earth. where access to the global information superhighway is so easy that even the most isolated research station in Antarctica, the most remote town in Africa, or the farmer in rural India can use it. This is Elon Musk's bold ambition for Starlink; it is not simply a pipe dream.

In 2015, Musk launched Starlink under the guise of SpaceX, an ambitious plan to set up a massive network of little low-Earth orbit (LEO) satellites to deliver worldwide broadband internet access. The idea that everyone should have access to the opportunities that the internet has to offer, regardless of geography, informs this ambitious goal, which is on a par with Musk's earlier endeavors.

This is a massive undertaking to build a satellite network. Significant constraints include technological difficulties, regulatory obstacles, and the significant expenditure needed. Musk didn't back down, though. Instead, he used the approaches that had been successful for him in his prior endeavors: iterative development, a dedication to in-house production, and an unwavering will to challenge the existing quo.

To understand the depth of the challenge, consider this: Traditional internet services rely on vast terrestrial infrastructures, fibre-optic cables that physically connect users to the internet. In many parts of the world, especially remote and rural regions, these infrastructures simply do not exist. With Starlink, Musk proposes a skyward shift, bypassing terrestrial limitations by beaming internet down from space.

By 2023, SpaceX had already launched over 1,500 Starlink satellites. This 'constellation', as it's referred to, is a game-changer. It could democratize internet access, fuel economic growth in developing nations, bridge the digital divide, and connect the unconnected, shaping a more equitable digital future. The potential socioeconomic impact is staggering, reflecting Musk's ability to dream big and think beyond traditional constraints.

However, there are legitimate concerns. While some astronomers raise concerns about the interference of "light pollution" with celestial studies, others raise concerns about the practicality of operating such a vast network of satellites. By lowering satellite brightness and developing the first automatic collision-avoidance systems, Musk has responded in typical Musk fashion: with incremental solutions and open communication.

Despite the challenges, it's hard not to be inspired by the sheer scale of Musk's vision. Starlink's success could radically redefine our perception of the internet, shifting it from a luxury to a basic human right. Moreover, Musk plans to use the profits from Starlink to fund his ultimate dream: Making life interplanetary.

Looking at Musk's journey, we're reminded of computer scientist Alan Kay's words, "The best way to predict the future is to invent it." Musk isn't waiting for the world to change; he's actively moulding it, propelling us towards a future that only existed in science fiction.

As we explore further, we must ask ourselves: What limitations are we accepting without question? How can we adopt Musk's big-thinking approach to overcome these barriers? How can we contribute to the creation of an inclusive, sustainable, and connected future?

Starlink, like all Musk's ventures, serves as a vivid illustration of his indomitable spirit. It reminds us that the impossible is only a temporary condition, a boundary waiting to be transcended. As we journey through the stars with Musk, let us bear in mind these lessons and let them inspire us to chase our own audacious dreams, no matter how unreachable they may seem.

In the words of Nelson Mandela, "It always seems impossible until it's done." May this sentiment guide us as we navigate the realm of possibilities, the kingdom of big thinkers. As we delve deeper into Musk's story, remember, the power of disruption lies not in replicating what has been done before, but in daring to envision what might be possible.

The undertaking of Starlink is not merely about providing faster internet or disrupting the telecommunications industry; it is about leveling the playing field. It is about ensuring that every child whether in the bustling city or remote countryside, has access to the same knowledge and the same opportunities. It underscores Musk's belief in an interconnected and equitable world.

Musk's approach to these enormous challenges offers a blueprint for anyone striving to bring about meaningful change. It's a testament to the power of iterative learning, relentless ambition, and above all, the audacity of dreaming big. This audacious dream is what sets Musk apart and what pushes him to transform the impossible into the possible.

Just as the stars in the night sky inspire wonder and ignite our imagination, so too do the innovations and ambitions of visionaries like Musk. They push the boundaries of what we believe is achievable, expanding our understanding of the world, and indeed, the universe we inhabit.

Musk's drive to connect the world is not only about technology or business success. It's about human connection, about breaking down barriers that hinder the sharing of ideas and knowledge. It is a reminder that regardless of where we come from, we are all residents of the same planet, inheritors of the same future.

To conclude, let's reflect on a profound thought from the legendary futurist and science fiction writer, Arthur C. Clarke: "The only way to discover the limits of the possible is to go beyond them into the impossible." With Starlink, Musk doesn't just push the boundaries of what we perceive as possible; he dissolves them, inviting us all to join him in this bold expedition into the future. As we delve deeper into the expansive canvas of Musk's visions and dreams, let's keep this audacious spirit at heart, for the impossible, after all, is often untried.

21. PayPal - Redefining Online Payments

BEFORE THE ADVENT of SpaceX and Tesla, before the audacious ambitions of Hyperloop and Starlink, Elon Musk made his mark on the world of technology and entrepreneurship through a disruptive financial services platform: PayPal. The story of PayPal offers a remarkable early glimpse into Musk's unique style of thinking big and his keen ability to anticipate the future and carve out innovative solutions for the forthcoming digital era.

In the late 1990s, the digital revolution was in its infancy. E-commerce was rapidly evolving, yet online payments remained a significant hurdle. Most transactions were dependent on traditional banking systems that were unfit for the burgeoning world of online commerce. Observing this bottleneck, Musk, with his forward-thinking prowess, envisioned an internet-native payment system. A solution that would democratize access to financial services, transcending geographic barriers, and banking bureaucracies.

In 1999, Musk founded X.com, an online bank, that after a series of acquisitions and strategic shifts would eventually become PayPal. The journey was fraught with challenges. Regulatory hurdles were substantial, as were the issues concerning fraud and security. Traditional financial institutions were skeptical and unyielding and often viewing X.com as a threat rather than an ally.

Nevertheless, Musk remained unfazed, adopting a disruptive mindset. He persisted with his goal which is redefining the way money moved around the world. He set up a team of ingenious engineers and nurtured a culture of relentless innovation - a strategy that would become a trademark of his future ventures.

Musk's big thinking behind PayPal was not merely about creating an online payment system. It was about empowering individuals and businesses by providing them access to global markets, thus driving social and economic development. Musk's vision for PayPal was transformative and its impact is unprecedented.

Fast forward to the present, PayPal is one of the world's largest online payment companies, boasting over 375 million active accounts by the end of 2022, according to the Wall Street Journal. It revolutionized the digital payments industry, fueling the growth of e-commerce and offering financial inclusion to millions who had been previously excluded from the traditional banking system.

PayPal was also a testament to Musk's remarkable foresight. As the New York Times observed, "Long before the iPhone, long before we were carrying our lives in our pockets, Musk imagined a world in which money moved as freely as information. He saw, far earlier than most, that financial systems were about to undergo an enormous change."

The story of PayPal underscores Musk's unique ability to identify gaps and visualize groundbreaking solutions that drive large-scale impact. As we delve into this tale of technological disruption, ask yourself, are there inefficiencies in your domain that you can address? Can you envision innovative solutions that disrupt the status quo and drive large-scale impact?

As we delve further into Musk's extraordinary life and achievements, remember, it all started with a desire to facilitate a basic human need – to transact, to exchange, and to progress. In Musk's world, no challenge is too big if it leads to a solution that changes the world for the better. As we chart the course of this remarkable innovator, let's imbibe this spirit, for it is within this audacity of vision where the seeds of the extraordinary are sown.

PayPal set a precedent for the entrepreneurial world; it demonstrated how audacious vision, coupled with relentless execution, can disrupt an entrenched industry. And it showed that in the digital age, borders and bureaucracies need not hinder human progress.

It's important to highlight that Musk didn't just stop at developing an online payment platform. He worked on transforming PayPal into a global brand that people could trust. He built partnerships, focused on superior customer service, and most importantly, delivered a product that simplified people's lives. It exemplified Musk's understanding of the need to humanize technology, to ensure it served its core purpose of enriching people's lives.

The legacy of PayPal extends beyond its services. The Economist noted that PayPal's alumni, often referred to as the 'PayPal Mafia,' includes many who have gone on to establish or lead other revolutionary tech companies. Among them are Peter Thiel, co-founder of Palantir; Reid Hoffman, co-founder of LinkedIn; and even Chad Hurley, Steve Chen, and Jawed Karim, who together created YouTube. This underlines the culture of innovation and big thinking that Musk fostered.

Despite selling PayPal to eBay in 2002, Musk's influence on the company and the world of digital finance remains. His pioneering work paved the way for the explosion of FinTech, inspiring a new generation of entrepreneurs to innovate within the financial industry.

Reflecting on Musk's journey with PayPal, it's evident how his big thinking helped shape the digital landscape. It shows us that disruption isn't just about technology, but about the transformative potential of ideas.

Musk's entrepreneurial journey, where his ventures are not mere businesses but solutions to real-world problems, is encapsulated by the thought-provoking statement made by Harvard Business Manager at the end of this chapter: "True innovators do not set out merely to innovate, but to solve problems, to make things better than they are, to disrupt the old way of doing things."

Musk's PayPal experience teaches us to defy convention, have the guts to think unconventionally, and have the boldness to act on those ideas. As we take this inspiration forward, let's keep in mind that change begins with an idea, but its realization requires action. The narrative of PayPal is not just about a company. It's about the embodiment of Musk's philosophy, of thinking big and disrupting the status quo. As we close this chapter, let's ponder over an interesting question – If an individual can democratize financial transactions.

22. Embracing the Impossible: Musk's Visionary Impact

ELON MUSK, a person who is now associated with big ideas and revolutionary inventions, frequently challenges the accepted knowledge that holds most of us behind. His life philosophy, "When something is important enough, you do it even if the odds are not in your favor," not only guides his work life but also informs how he sees the world. Musk is a visionary with a remarkable talent for recognizing potential where others see impossibilities, and he has consistently turned far-fetched concepts into businesses that have changed the world.

SpaceX, Musk's ambitious aerospace manufacturer and space transportation company, exemplifies this ethos. When he announced his intentions to make space travel more affordable and eventually colonize Mars, the idea was derided by industry veterans and observers alike. Many considered it an impossible endeavor fraught with insurmountable technical and financial challenges.

However, Musk had a different perspective. He saw it as an opportunity to alter the history of humanity and take steps to ensure our existence after Earth. In the face of strong skepticism, Musk started from scratch to create SpaceX. Every failure was viewed as a chance to grow and learn, and every setback as a step closer to success.

Despite the initial dismissive attitudes, SpaceX achieved the unthinkable. In 2012, it became the first privately-funded company to send a spacecraft, the Dragon, to the International Space Station. The successful reuse of Falcon 9's first stage in 2017 marked another significant milestone, substantially reducing the cost of space exploration. The world watched in awe as Musk's 'impossible' dreams began to manifest into reality. Musk's audacious thinking not only shifted the paradigm of space travel but also ignited a renewed public interest in space exploration.

Musk's ability to think big extends beyond SpaceX. Tesla, his electric car company, had a similarly transformative effect on the automotive industry. At a time when electric vehicles were an afterthought, Musk envisioned a sustainable transportation future. Undeterred by the many obstacles, from production bottlenecks to financial concerns, Tesla persevered. Today, Tesla's innovative electric vehicles and energy solutions have pushed the entire industry towards a greener future, causing a seismic shift in the way we perceive mobility.

Musk's thought process represents a break from traditional problem-solving approaches. Rather than improving existing systems incrementally, Musk employs 'first principles thinking', a strategy often attributed to physicists. This approach involves breaking down a problem to its fundamental principles and then reconstructing a solution from scratch. This methodology has underpinned Musk's most revolutionary ventures, such as SpaceX's reusable rockets and Tesla's electric vehicles.

Despite the criticisms and the many hurdles, Musk's willingness to embrace the impossible has not only revolutionized industries but also inspired countless others to think big. His story sends a powerful message about the transformative power of unyielding belief in one's vision.

In analyzing Musk's approach, there's a question worth pondering: Are we limiting our potential by adhering to conventions and not daring to dream big? If a single person's audacious dreams could reshape industries and propel humanity towards a more sustainable and interplanetary future, what could we achieve collectively if we dared to embrace the impossible?

Elon Musk's path is evidence of this, showing that the seemingly impossible can be overcome with daring thought and unwavering persistence. The real change isn't only in the sectors Musk has upended; it's also in the minds he has propelled to "think big."

Take Neuralink, Musk's bold venture into the field of neurotechnology. This project aims to merge the human brain with artificial intelligence, an idea once confined to the realms of science fiction. The potential applications are breathtaking, ranging from curing neurodegenerative diseases to enhancing human cognition. Many question the feasibility of such a venture, but for Musk, this is just another challenge awaiting his touch of innovation. Just as he disrupted the aerospace and automotive industries, Musk aims to revolutionize our understanding of the human brain and its potential.

Musk has maintained his vision and drive throughout his trip. He thinks that accepting setbacks is an essential component of success. This attitude has been important in making his bold ideas a reality, along with a persistent work ethic and a talent for upending the status quo.

In Musk's world, thinking big isn't a fanciful exercise; it's a way of life, a mindset that disregards perceived limits and continuously pushes the boundaries of possibility. This ethos has been a driving force behind Musk's game-changing ventures and is a critical lesson for future innovators and entrepreneurs.

Musk's ability to dream big and embrace the impossible is inspiring. It's a powerful reminder of human potential when we dare to think beyond our current limitations and pursue transformative visions. The impact of his ventures extends far beyond their respective industries, serving as beacons of inspiration for future generations of innovators and change-makers.

Reflecting on Musk's journey raises a thought-provoking question: In a world that's quick to label ideas as impossible, how can we foster a culture that not only encourages but also celebrates audacious thinking? As we look to the future, this question is worth pondering. If we are to tackle the most pressing challenges of our time, we need more dreamers, more innovators, more 'Elon Musks'.

As we conclude this chapter, we can take a leaf out of Musk's book: dreaming big isn't merely about setting lofty goals; it's about having the courage to challenge the status quo, the resilience to overcome adversity, and the unwavering conviction in the power of one's vision. In the words of Robert Browning, "Ah, but a man's reach should exceed his grasp, or what's a heaven for?"

Imagine what we could achieve as a society if we adopted Musk's audacious approach to thinking big. Imagine the innovations we could bring to life, the challenges we could overcome, and the future we could build. As we journey through Musk's life in this biography, let's challenge ourselves to question our assumptions, to defy the conventional, and to dream big, always remembering that the impossible is often the untried.

PART 3: GETTING THINGS DONE

23. The Power of 'First Principles' Thinking

ELON MUSK'S philosophy towards achieving monumental feats in diverse sectors from space exploration to renewable energy is hinged upon a profound cognitive strategy he refers to as 'First Principles' thinking. What does this strategy entail, and how does it underpin Musk's success across his array of groundbreaking enterprises?

Ancient philosophers like Aristotle and Descartes prominently supported the use of first principles thinking as a technique of research. In essence, it entails dissecting complicated issues into their underlying elements or underlying facts, then reconstructing solutions from scratch. It is a significant divergence from conventional approaches, which frequently depend on preexisting norms, parallels, or precedents.

Musk provides a vivid illustration of this strategy when discussing the early days of SpaceX. At that time, the costs of purchasing a rocket were prohibitively high. Instead of succumbing to these market constraints, Musk chose to apply First Principles thinking. He identified the raw materials that go into making a rocket, calculated their costs, and concluded that he could build a more affordable rocket from scratch. This breakthrough insight birthed SpaceX, proving that space travel need not be a privilege for the affluent few.

In Tesla's context, First Principles thinking enabled Musk to challenge the automotive industry's conventional wisdom. While others argued electric cars couldn't compete with gas-powered vehicles due to battery cost, Musk focused on the fundamental capabilities of lithium-ion cells. By innovating around battery design, manufacturing, and supply chain management, Tesla delivered electric cars that could compete on performance and price, disrupting the automobile industry's status quo.

First Principles thinking is more than a problem-solving strategy; it's a mindset that embraces curiosity, a willingness to question established norms, and the courage to innovate. It requires us to discard our preconceived notions and biases, and to see the world through a more fundamental, objective lens.

The application of this mindset has not been without challenges. Musk has faced skepticism and adversity, but he stood unwavering in his vision. Armed with First Principles thinking, he tackled each obstacle, learning and iterating along the way. This approach has driven remarkable economic results across his ventures, with SpaceX revolutionizing the economics of space travel and Tesla becoming a frontrunner in the electric vehicle market.

Reflecting on the transformative impact of First Principles thinking in Musk's journey prompts us to question: How often do we limit ourselves by accepting the prevailing assumptions in our fields? Can we reshape our industries, as Musk did, by adopting a First Principles mindset?

Applying this mode of thinking, as Musk demonstrates, isn't confined to launching rockets or manufacturing electric cars. It can apply to any area of life where complex challenges reside and innovative solutions are required.

Musk's unique cognitive style, amalgamated with a relentless pursuit of ambitious goals, represents a potent formula for 'getting things done'. It underlines a fundamental truth about human potential: with the right mindset and an audacious vision, we can challenge the status quo and drive disruptive innovation.

As we reflect on Musk's journey, let's challenge ourselves to question prevailing assumptions, to embrace First Principles thinking, and to drive innovation in our respective fields. Remember, the spark of innovation doesn't reside solely within the realm of tech billionaires, but in the mind of every individual daring to ask, 'Why not?'.

Despite the unparalleled success Musk has achieved with First Principles thinking, it's important to recognize that it isn't a panacea for all problems. It's an intense, demanding process that requires a deep understanding of a problem, a willingness to question everything, and the courage to defy conventional wisdom. The successes of SpaceX and Tesla didn't come easy; they were born out of years of relentless hard work, countless failures, and a tenacious spirit that refused to surrender to adversity.

Musk's use of First Principles thinking also extends to his management style. He promotes a culture of open communication, encouraging his employees to voice their ideas and criticisms freely, no matter their position in the company hierarchy. This approach not only fosters innovation but also ensures that decisions are made based on their merit and not on who proposes them. By valuing the input of every team member, Musk demonstrates a leadership style rooted in the same principles of logic and reasoning that guide his decision-making process.

The bold application of First Principles thinking in his ventures, coupled with his progressive leadership style, has earned Musk admiration and respect from his peers and competitors alike. He has shown that this mode of thinking, when paired with a daring vision and a resolute spirit, can overcome even the most formidable challenges. His journey serves as a powerful reminder that conventional wisdom can and should be challenged if we are to make significant strides forward.

As we reflect on the insights garnered from Musk's journey, we must ask ourselves: How can we incorporate First Principles thinking into our own lives and careers? How can we cultivate the courage to challenge the status quo, rethink accepted norms, and push the boundaries of what's considered possible?

In conclusion, Elon Musk's extraordinary journey teaches us that getting things done requires more than just ambition and hard work. It involves a mindset shift, a willingness to question, to learn, and to adapt. The application of First Principles thinking, despite its inherent challenges, can be a game-changer in problem-solving and innovation.

Drawing from Musk's own words, let's remind ourselves: "You get paid by doing or making something people want, and those who make change are those who get ahead." As we navigate the complex landscapes of our respective fields, let's dare to think differently, to challenge the norms, and to make change happen.

24. Bold Decision Making: The Art of Calculated Risk-Taking

Diving into Elon Musk's journey reveals a leader who embraces calculated risks. The audacious ventures of SpaceX, Tesla, SolarCity, and the Boring Company are testaments to this philosophy. But how does Musk approach risk-taking? Is it a reckless leap of faith or a meticulous calculation grounded in science, logic, and entrepreneurial intuition?

In Musk's entrepreneurial narrative, the decision to form SpaceX, an audacious endeavor aiming to commercialize space travel, embodies his propensity for calculated risk-taking. During SpaceX's early days, Musk was warned repeatedly about the high failure rate of rocket startups. Despite this, he ventured forth, injecting his personal capital and betting against the odds.

But his decision wasn't a reckless gamble. It was born from an understanding of the fundamentals of rocket science, a belief in his team, and a passionate conviction that space exploration was critical for humanity's future. He understood the risks but also saw an opportunity too compelling to pass up.

Similarly, when Musk invested in Tesla, electric automobiles were thought to be a pipe dream, incapable of competing with internal combustion engines. Rather than trusting popular knowledge, Musk investigated the underlying principles of energy and transportation. He recognized a technical change in battery technology and increased environmental awareness as indicators of the possibility for upheaval in the car sector. He saw the possible rewards outweighing the huge hazards once more, and he accepted the gamble.

Musk's calculated risk-taking approach also translates into his leadership and decision-making processes. He's been known to set aggressive timelines and ambitious targets. Take, for instance, his goal to have a million robo-taxis on the road by 2020, an objective many industry experts considered outlandish. However, in Musk's philosophy, setting ambitious goals drives innovation and progress even if it means falling short occasionally.

It's important to note that while Musk is bold, he's not impulsive. His decisions no matter how daring they appear are based on data, research, and reasoning. This meticulous and analytical approach has, over time, turned seemingly impossible ideas into reality and yielded substantial economic results. Tesla's market cap surpassing that of traditional car manufacturers and SpaceX's breakthroughs in reusable rockets stand as clear evidence of the success of this approach.

Taking chances is a difficult task. It opens the door to failure, criticism, and loss. However, Musk's experience tells us that taking a certain degree of risk is not only beneficial—it is required—in order to innovate and achieve great gains. It is not about jumping into the unknown blindly, but rather about doing your research, comprehending the landscape, balancing the benefits and drawbacks, and making an informed decision.

To quote Musk himself: "When something is important enough, you do it even if the odds are not in your favor." This bold mindset is a call to action for all who aspire to make a difference. It implores us to embrace calculated risk-taking, to dare to disrupt, and to have the courage to turn audacious ideas into reality.

Are you willing to take calculated risks to achieve what's important to you? The next chapter in innovation and disruption awaits those bold enough to write it. Are you ready to pick up the pen?

In the context of Elon Musk's companies, this approach to risk-taking has led to groundbreaking innovations. From electric vehicles achieving mainstream acceptance to rockets landing back on Earth that are ready to be reused. These monumental achievements were initially met with skepticism, if not outright disbelief. But Musk's ability to identify the potential rewards amidst considerable risks created a pathway for these disruptive ideas to become realities.

This audacious, deliberate attitude to risk-taking is at the heart of the dynamic culture of invention cultivated in Musk's businesses. These settings promote the collision of ideas, which fosters creativity and occasionally results in novel solutions to challenging issues. Unquestionably, a key factor in the organization's ongoing success is Musk's ability to embed this ethos inside them.

But the journey hasn't been without its hardships. Each of Musk's companies has faced moments of existential crisis. SpaceX nearly went bankrupt after three failed launches. Tesla was only days away from bankruptcy during the 2008 financial crisis. But in these critical moments, Musk didn't retreat. Instead, he doubled down, investing more of his personal fortune into the companies he so fervently believed in. It was his belief in the core principles of his ventures, despite the financial risk, that carried them through the storm.

These instances offer profound insights into Musk's character. They portray a man who is not only a visionary but also a tenacious fighter who stands unwavering in the face of adversity. As Musk himself said, "If you get up in the morning and think the future is going to be better, it is a bright day. Otherwise, it's not."

Reflecting on Musk's approach to risk-taking invites us to consider our own attitudes towards risk and failure. Are we letting fear of failure inhibit our progress? Or are we bold enough to take calculated risks, even in the face of uncertainty and skepticism?

Musk's methodology emphasizes the importance of critical thinking and objective analysis in decision-making. It urges us to question the status quo and not to shy away from challenging established norms. For innovators and disruptors charting their own courses, adopting this approach could make the difference between playing it safe and making breakthrough decisions.

In conclusion, Musk's approach to calibrated risk-taking provides insightful advice for every ambitious businessperson or leader. His experience serves as an inspiring reminder that making a bold foray into the unknown area while armed with information, conviction, and tenacity may produce amazing outcomes. Let's look at problems as chances for development and creativity rather than as insurmountable impediments. Because the ability to influence the future lies with people who dare to think outside the box and take calculated risks.

25. Embracing Failure: The Ultimate Learning Curve

A FAMED FUTURIST and billionaire industrialist, Elon Musk's successes are indisputable. From SpaceX to Tesla, his endeavors have disrupted industries and reshaped our perception of what's possible. Yet, it's not just his victories that define him. It's the failures he's weathered, the resilience he's displayed, and the lessons he's absorbed from each setback that have truly shaped Musk and his ventures.

In an era where failure is often stigmatized, Musk stands as a paragon of a different perspective—one that views failures not as dead-ends but as opportunities for learning, growth, and eventual triumph.

One can trace this attitude back to the nascent days of SpaceX. Musk's dream of making space exploration affordable was on the brink of collapse after three failed rocket launches. Many, if not most, would have cut their losses and abandoned the endeavor. Yet Musk persisted, learning from each failure and adjusting his strategy. With the fourth launch, success arrived in the form of a contract with NASA, and the rest, as they say, is history.

This resilience in the face of failure isn't limited to SpaceX. When Tesla was grappling with production issues and delays in the Model 3 rollout, instead of passing the buck, Musk took personal responsibility. He worked on the factory floor, personally overseeing solutions to problems. The episode served as a testament to Musk's ability to not only endure failure but to actively engage with it, learn from it, and transform it into a stepping stone to success.

And therein lies the crux of Musk's approach towards failure—it's about embracing it as an integral part of the innovation process. He famously said, "Failure is an option here. If things are not failing, you are not innovating enough."

Musk's attitude towards failure extends beyond words; it penetrates the corporate cultures of his businesses, creating an atmosphere that celebrates risk-taking, stimulates innovation, and sees failure as a chance to grow. His enterprises have been able to advance innovation by always pushing the envelope thanks to this mindset.

But what can we learn from Musk's approach to failure? For one, it encourages us to shift our perspective on failure, viewing it not as a threat but as an ally on our path to success. It also highlights the importance of resilience and tenacity in the face of adversity—qualities that are integral to any successful entrepreneurial venture.

Furthermore, it underscores the critical role of learning in the innovation process. Each failure offers valuable insights and lessons that can guide future decisions and strategies, ultimately leading to success. As Musk aptly puts it, "When you struggle with a problem, that's when you understand it."

As we journey through the unpredictable landscape of innovation and entrepreneurship, let us take to heart Musk's philosophy of embracing failure. As we stumble, may we remember that each fall is but a step on the path to success. And in every moment of doubt, let us recall Musk's words: "If something is important enough, you should try, even if the probable outcome is failure."

Finally, as we forge our paths, we should contemplate this: How can we reshape our perception of failure, not as an obstacle to be feared, but as a valuable teacher on our journey to success? The answer to this question could very well change the trajectory of our journey, propelling us towards success in the face of adversity.

We learn from Musk's epic journey that accepting failure—the ultimate learning curve—is intrinsically bound up with his success. Musk's strategy goes beyond simple perseverance; it deliberately examines failure, draws out important lessons, and then applies them to new difficulties.

In 2008, it appeared as though Musk's world was about to end. SpaceX still hadn't managed a successful launch, and Tesla was only days away from filing for bankruptcy. It was a time marked by difficulties and failures, illustrative of the frequently perilous route of invention. But Musk persisted rather than giving up. His faith in them, determination, and capacity to learn from them brought both businesses back from the edge and put them on the road to leadership in their respective industries.

During this period, Musk learned invaluable lessons about leadership, logistics, and the value of perseverance—lessons that he continues to apply in his ventures. In the following years, we've seen SpaceX revolutionize space travel and Tesla emerge as a leader in the electric vehicle industry. This transformation wasn't a mere stroke of luck; it was the result of Musk's ability to learn from failure, adjust his strategies, and forge ahead.

Musk's journey gives us insights into how failure, often perceived as the antithesis of success, can instead be its progenitor. His story serves as a blueprint for entrepreneurs and innovators to not just endure failures but to welcome them as opportunities for growth and learning. After all, in the laboratory of innovation, failure is not the end, but a vital step in the process of discovery.

As we navigate our own paths, the question we must continually ask ourselves is: How can we turn our setbacks into comebacks?

How can we reframe our failures as stepping stones to success? In the face of adversity, are we willing to learn, adapt, and keep pressing forward?

Embracing failure as Musk does requires a shift in mindset—one that views failure not as a symbol of defeat, but as a wellspring of knowledge. It's about understanding that setbacks are often the universe's way of redirecting us towards a new path, one that leads us closer to our goals.

So, as we reach the end of this chapter, let's reflect on the lessons learned from Musk's approach to failure. How can we imbibe his resilient spirit? How can we welcome failure as an integral part of our journey rather than a detour? And, most importantly, how can we leverage our failures as a springboard towards greater success?

As Thomas Edison once said, "I have not failed. I've just found 10,000 ways that won't work." In the spirit of Edison and Musk, let's celebrate our failures, not because they're pleasant, but because they are our most effective teachers, guiding us towards our next big success. Are you ready to embrace failure, learn from it, and use it to fuel your journey towards success? After all, it is often through the crucible of failure that great success is forged.

26. Transcending Boundaries: The Multidisciplinary Approach to Problem-Solving

At the heart of Musk's problem-solving approach lies an unwavering commitment to multidisciplinary thinking, the cross-pollination of ideas across different fields. This chapter, "Transcending Boundaries: The Multidisciplinary Approach to Problem-Solving," examines Musk's knack for merging disparate disciplines into a cohesive framework for innovation.

Think about Tesla's breakthroughs in battery technology to see how Musk uses this strategy. Musk combined insights from physics, chemistry, and material science rather than just relying on automobile engineering. Musk advanced battery research by going beyond conventional limitations, propelling Tesla to the top of the electric vehicle market. This sums up Musk's interdisciplinary mindset: fusing several fields of expertise to develop and address challenging issues.

This strategy is demonstrated by Musk's endeavors, including SpaceX and Neuralink. Each business embodies the blending of several fields, as well as science, technology, and a bold future vision. They serve as an example of how combining many disciplines may unleash incredible innovation and make the seemingly impossible attainable.

But how does one cultivate such a mindset? It begins by developing an insatiable curiosity, a love for learning that spans multiple fields. Musk's extensive knowledge of physics, engineering, artificial intelligence, and economics wasn't obtained overnight. It is the result of years of self-learning, reading, and questioning of consistently striving to understand the world from various perspectives.

The capacity to see connections when others miss them is another crucial aspect. Musk doesn't see separate things when he examines an electric vehicle, a rocket, or a high-speed transit system. He observes the blending of several components, each of which stands for a distinct body of knowledge. A distinguishing feature of Musk's thinking is his ability to make connections and to perceive the whole rather than merely the bits.

By adopting this multidisciplinary approach - we, too, can enhance our problem-solving capabilities. We can learn to think beyond our area of expertise, to draw connections across different fields, and to see the world not as a series of isolated domains but as an interconnected web of knowledge. This mindset invites us to question, explore, and venture into uncharted territories of thought and innovation. It challenges us to continually expand our understanding and transcend the boundaries of our knowledge.

So, as we delve into this complex yet rewarding world of multidisciplinary thinking, let's reflect on its implications. How can we apply this approach to our own lives and careers? How can we cultivate a mindset that constantly seeks to learn and draw connections across different fields? And most importantly, how can we leverage this mindset to drive innovation, solve complex problems, and make a positive impact in the world?

Reflecting on the words of Steve Jobs, "Creativity is just connecting things," let's remember that the ability to connect different fields is a powerful catalyst for innovation. Musk's journey serves as a testament to this. It challenges us to think beyond our comfort zones, to embrace the power of multidisciplinary thinking, and to become architects of the future.

"Transcending Boundaries: The Multidisciplinary Approach to Problem-Solving," illuminates a significant aspect of Elon Musk's methodology to innovation. It unveils how, rather than being constrained by conventional wisdom and traditional boundaries. Musk thinks broadly, drawing from an array of disciplines to tackle intricate problems. Yet, this is just the start of understanding Musk's extraordinary aptitude for "getting things done."

Elon Musk's leadership approach at SpaceX offers a compelling case study of his multidisciplinary mindset in action. SpaceX's goal is not simply to build rockets; it's about transforming space travel to make life multiplanetary, a vision that demands a convergence of expertise in rocket science, artificial intelligence, metallurgy, and many other areas. When faced with the formidable challenge of reducing the cost of space travel, Musk didn't confine himself to the constraints of existing aerospace engineering norms. Instead, he approached the issue from a fundamental physics perspective, inquiring about the raw materials cost of a rocket. The discrepancy between this cost and the market price of rockets led him to conclude that there must be inefficiencies in the production process, a problem he attacked from multiple angles resulting in the development of reusable rockets and a feat deemed unattainable by traditional aerospace entities.

Further evidence of Musk's multidisciplinary approach can be found in his ambitious venture, Neuralink. This bold undertaking merges neuroscience, robotics, and artificial intelligence with an audacious vision of the future: a high-bandwidth interface between the human brain and machines. Musk's capability to synthesize knowledge across these disciplines is instrumental to the groundbreaking work at Neuralink. It stands as another testament to his unique ability to transcend the conventional boundaries of knowledge and create solutions that redefine possibilities.

Musk's multidisciplinary thinking is not an innate talent, but a cultivated skill. It starts with curiosity and a refusal to be confined by the rigid delineations between academic disciplines. It requires a commitment to lifelong learning, an openness to new ideas, and the courage to disrupt established conventions. It involves seeing the world not merely as fragments but as an interconnected network of knowledge, where solutions emerge from the fusion of different fields.

By fostering this multidisciplinary mindset, we too can enhance our capacity to think innovatively and solve complex problems. It encourages us to look beyond our existing knowledge, to connect disparate ideas, and to see possibilities where others see limits. As Steve Jobs once said, "Innovation is what distinguishes between a leader and a follower." By embracing Musk's approach to multidisciplinary thinking, we can be leaders in our own fields, carving out new paths and propelling innovation forward.

Let's finish with a remark from Albert Einstein that captures the very core of Musk's interdisciplinary approach: "The significant problems we face cannot be solved at the same level of thinking we were at when we created them." It is an invitation to think more critically, to see past divisions, and to value the strength of a broad knowledge base.

27. The Work Ethic Marathon: Musk's Productivity Secret

An uncompromising dedication to interdisciplinary thinking, or the cross-pollination of ideas across multiple domains, sits at the core of Musk's problem-solving methodology. In the chapter "Transcending Boundaries: The Multidisciplinary Approach to Problem-Solving," Musk's talent for fusing many disciplines into a unified framework for creativity is examined.

Think about Tesla's breakthroughs in battery technology to see how Musk uses this strategy. Musk combined insights from physics, chemistry, and material science rather than just relying on automobile engineering. Musk advanced battery research by going beyond conventional limitations, propelling Tesla to the top of the electric vehicle market. This sums up Musk's interdisciplinary mindset: fusing several fields of expertise to develop and address challenging issues.

Musk's ventures, from SpaceX to Neuralink, are a testament to this approach. Each company represents a fusion of disciplines, a blend of technology, science, and an audacious vision for the future. They exemplify how merging different fields can unlock unprecedented innovation, making the seemingly impossible, possible.

But how does one cultivate such a mindset? It begins by developing an insatiable curiosity, a love for learning that spans multiple fields. Musk's extensive knowledge of physics, engineering, artificial intelligence, and economics wasn't obtained overnight. It is the result of years of self-learning, reading, and questioning, of consistently striving to understand the world from various perspectives.

Another critical factor is the ability to see connections where others don't. When Musk looks at an electric car, a rocket, or a high-speed transportation system, he doesn't see isolated entities. He sees the fusion of different elements, each representing a domain of knowledge. This capacity to draw connections, to see the whole rather than just the parts, is a hallmark of Musk's thinking.

By adopting this multidisciplinary approach, we, too, can enhance our problem-solving capabilities. We can learn to think beyond our area of expertise, to draw connections across different fields, to see the world not as a series of isolated domains but as an interconnected web of knowledge. This mindset invites us to question, explore, and venture into uncharted territories of thought and innovation. It challenges us to continually expand our understanding and transcend the boundaries of our knowledge.

As we delve into this complex yet rewarding world of multidisciplinary thinking, let's reflect on its implications. How can we apply this approach to our own lives and careers? How can we cultivate a mindset that constantly seeks to learn and draw connections across different fields? And most importantly, how can we leverage this mindset to drive innovation, solve complex problems, and make a positive impact in the world?

Reflecting on the words of Steve Jobs, "Creativity is just connecting things," let's remember that the ability to connect different fields is a powerful catalyst for innovation. Musk's journey serves as a testament to this. It challenges us to think beyond our comfort zones, to embrace the power of multidisciplinary thinking, and to become architects of the future.

"Transcending Boundaries: The Multidisciplinary Approach to Problem-Solving," illuminates a significant aspect of Elon Musk's methodology to innovation. It unveils how, rather than being constrained by conventional wisdom and traditional boundaries, Musk thinks broadly, drawing from an array of disciplines to tackle intricate problems. Yet, this is just the start of understanding Musk's extraordinary aptitude for "getting things done."

Elon Musk's leadership approach at SpaceX offers a compelling case study of his multidisciplinary mindset in action. SpaceX's goal is not simply to build rockets; it's about transforming space travel to make life multiplanetary, a vision that demands a convergence of expertise in rocket science, artificial intelligence, metallurgy, and many other areas. When faced with the formidable challenge of reducing the cost of space travel, Musk didn't confine himself to the constraints of existing aerospace engineering norms. Instead, he approached the issue from a fundamental physics perspective, inquiring about the raw materials cost of a rocket. The discrepancy between this cost and the market price of rockets led him to conclude that there must be inefficiencies in the production process, a problem he attacked from multiple angles, resulting in the development of reusable rockets, and a feat deemed unattainable by traditional aerospace entities.

Further evidence of Musk's multidisciplinary approach can be found in his ambitious venture, Neuralink. This bold undertaking merges neuroscience, robotics, and artificial intelligence with an audacious vision of the future: a high-bandwidth interface between the human brain and machines. Musk's capability to synthesize knowledge across these disciplines is instrumental to the groundbreaking work at Neuralink. It stands as yet another testament to his unique ability to transcend the conventional boundaries of knowledge and create solutions that redefine possibilities.

Musk's multidisciplinary thinking is not an innate talent, but a cultivated skill. It starts with curiosity and a refusal to be confined by the rigid delineations between academic disciplines. It requires a commitment to lifelong learning, an openness to new ideas, and the courage to disrupt established conventions. It involves seeing the world not merely as fragments but as an interconnected network of knowledge, where solutions emerge from the fusion of different fields.

By fostering this multidisciplinary mindset, we too can enhance our capacity to think innovatively and solve complex problems. It encourages us to look beyond our existing knowledge, to connect disparate ideas, and to see possibilities where others see limits. As Steve Jobs once said, "Innovation is what distinguishes between a leader and a follower." By embracing Musk's approach to multidisciplinary thinking, we can be leaders in our own fields, carving out new paths and propelling innovation forward.

Taking the lessons from Musk's work ethic, we delve deeper into his multifaceted approach to tackling challenges. As well as sheer grit, it's critical to recognize the role of strategic thinking in his toolkit. Musk has perfected the art of problem decomposition, segmenting larger challenges into manageable parts, ensuring that no problem is insurmountable. This practice, combined with his marathon-like stamina, facilitates a productive approach to every hurdle that comes his way.

Additionally, Musk consistently emphasizes the need for feedback and critical thinking. He often challenges his team, encouraging them to critique his ideas, thereby fostering a culture of innovation and iterative improvements. He practices what he terms as 'radical candor,' creating an environment that values honesty and directness over hierarchical structure.

Musk's work ethic extends to his learning strategies too. He is known for a technique dubbed 'learning transfer', where he applies knowledge from one context to another. For example, his understanding of physics principles enabled him to grasp complex aerospace concepts, crucial in leading SpaceX's endeavors.

Moreover, Musk exhibits an exceptional ability to prioritize and focus. He zeroes in on what matters most and ruthlessly eliminates or delegates the rest. This level of focus might seem extreme to some, but for Musk, it's a strategy integral to juggling multiple high-stake projects simultaneously.

However, no analysis of Musk's work ethic would be complete without discussing his resilience. He has endured numerous professional setbacks - from the brink of bankruptcy with Tesla in 2008 to multiple rocket launch failures at SpaceX. However, his indomitable spirit prevailed each time. In the face of adversity, he famously remarked, "If you get up in the morning and think the future is going to be better, it is a bright day. Otherwise, it's not."

Yet, while Musk's work ethic is undoubtedly a marvel, it's essential to remember that it's not the only path to success. His approach is unique and tailored to his personality, capacity, and the industries in which he operates. Each entrepreneur must find a rhythm and approach that works best for them and their ventures.

To conclude, let's consider a quote from Albert Einstein: "The significant problems we face cannot be solved at the same level of thinking we were at when we created them." This speaks to the very essence of Musk's multidisciplinary approach. It's a call to raise our level of thinking, to transcend boundaries, and to embrace the power of a diverse knowledge base. Albert Einstein once said, "It's not that I'm so smart, it's just that I stay with problems longer." This quote embodies the essence of Musk's work ethic, a relentless determination that continues to shape our future.

28. The Musk Vision: Thinking in Scale and Time

A PIVOTAL ELEMENT in understanding Elon Musk's staggering accomplishments is appreciating his unique perspective of scale and time. Elon Musk does not simply think big; he thinks in proportions that are sometimes hard for many of us to fully grasp.

One primary example is his long-term vision for SpaceX. Most of us might think a few years ahead, but Musk is strategizing for a time when Earth might no longer be habitable. He has publicly stated his goal to build a city on Mars by 2050, an audacious plan that surpasses any conventional business strategy. In Musk's view, a multi-planetary existence is not a fanciful concept but a practical and necessary goal for human survival.

Simultaneously, Musk applies a large-scale approach to problem-solving. Rather than being intimidated by the enormity of the problems he faces, he seems to thrive on the challenges they present. This is evident in his relentless pursuit of revolutionizing transportation systems on Earth through ambitious projects like the Hyperloop and The Boring Company. He sees not only the immediate, but the expansive impact these projects could have on our future.

It is this larger-than-life perspective that has enabled Musk to set extraordinary goals and push the boundaries of what is possible.

Whether it's launching reusable rockets, envisioning self-driving cars, or pioneering the integration of artificial intelligence with human cognition through Neuralink, Musk's view of scale and time differentiates him from many of his peers.

This method of thinking isn't only about having bravery or ambition. It's evidence of Musk's extraordinary capacity for dreaming and his unwavering conviction that those visions can be realized, no matter how long the process takes or how large the project. It involves having the patience and determination to see ideas that are bigger than oneself through.

However, thinking on such a grand scale can be intimidating. How can we apply Musk's principles to our lives and our businesses without feeling overwhelmed? It's about starting small, setting achievable goals, and gradually working towards something bigger. You don't have to aim for Mars, but you can adopt Musk's 'First Principle Thinking,' breaking down complex problems to their fundamental truths and reasoning up from there. This approach can help any budding entrepreneur navigate the path to innovation.

As French writer Antoine de Saint-Exupéry once said, "As for the future, your task is not to foresee it, but to enable it." This, in essence, encapsulates Musk's vision. As we move forward, how can we enable the future? What large-scale, long-term goals can we set for ourselves? And most importantly, how can we turn these goals into a reality? The answers to these questions may just define our journey towards disruptive innovation and progress.

Elon Musk's approach to scale and time is not just about ambition—it's also about the meticulous planning, execution, and perseverance that go behind each of his ventures. Whether it's his 12-hour workdays, his attention to detail, or his persistence in the face of failure, Musk is living proof that big dreams require big commitments.

Think about Tesla's journey. Starting from scratch in an industry dominated by century-old giants, Musk envisioned a world of sustainable transport. He aimed not just to create an electric car but to change the perception of what an electric car could be. A look at Tesla's "Master Plan," posted on their blog in 2006, showcases Musk's clear vision and long-term thinking. It outlined a clear four-part strategy, which has been largely followed over the past decade and a half.

In Musk's vision for SpaceX, we see his distinctive blend of ambition and long-term planning. His ultimate goal? Make life multiplanetary and build a self-sustaining city on Mars. Musk is not just dreaming of this future—he's meticulously planning for it. SpaceX's Starship, once fully operational, is planned to carry up to 100 people per flight to Mars. The timeline Musk envisions is bold —he hopes to start making regular flights by the mid-2020s and establish a human colony by 2050.

Musk's approach can teach us several things. To start, the importance of long-term planning cannot be overstated. The world's most significant achievements didn't happen overnight—they were the culmination of years, sometimes decades, of meticulous planning and execution.

Secondly, Musk's mindset underscores the power of thinking on a larger scale—not just in terms of profit or market share, but in terms of impact. How can our actions today shape the world of tomorrow? How can we create a future that is not just successful but sustainable?

Lastly, Musk's vision serves as a reminder of the power of perseverance. He has faced numerous setbacks along the way—failed rocket launches, production issues at Tesla, public skepticism, and more. Yet, he has never given up on his vision.

Let's ponder the words of theoretical physicist Albert Einstein: "You never fail until you stop trying." Musk embodies this spirit, and through his journey, he teaches us that to truly make a difference and create a sustainable future, we must be prepared to think big, plan long-term, and never give up.

As we wrap off this chapter, keep in mind how unique Musk's concept is. He looks at the future through a prism that emphasizes possibilities rather than issues. He teaches us that having large objectives doesn't only mean knowing the long-term effects of our choices and utilizing that information to inform our plans and judgement's.

29. Talent Magnet: Sourcing and Nurturing the Best Minds

ELON MUSK HAS FREQUENTLY BEEN DESCRIBED as a human magnet because of his capacity to draw the best talent to himself in whatever industry he enters. This is no accident. He previously remarked, "Talent is extremely important. It's like a sports team. The team that has the best individual player will often win, but then there's a multiplier from how those players work together and the strategy they employ."

The foundation of Musk's ability to not just draw in the brightest brains, but also keep them around and inspire them has been this idea. Let's examine how he did it at Tesla, SpaceX, Neuralink, and his other businesses to understand how he did it.

Firstly, his mission-driven approach sets him apart. Musk's companies aren't just businesses; they're crusades with grand, world-changing visions - whether it's reducing carbon emissions, making humans a multi-planetary species, or merging human consciousness with AI. It's easy to imagine how this could attract individuals with the same fervor and zest for changing the world.

John Chambers, former CEO of Cisco, in his interview with the Wall Street Journal, said, "Musk is offering something bigger than a paycheck. He's offering a purpose." When work transcends into a purpose, retention becomes easier. The 'mission' becomes the glue that holds the talent together, even in the face of adversity.

Secondly, Musk's leadership style is challenging yet empowering. He's known for setting seemingly impossible goals - referred to as 'stretch goals' in management parlance - but he also empowers his employees to meet these goals. He offers them autonomy, resources, and support, as well as the assurance that their leader is right there with them doing the work.

Musk's former assistant, Mary Beth Brown, shared in a New York Times interview, "He demands a lot, but he'll move mountains to ensure we have what we need to deliver." This environment breeds resilience, fosters innovation, and ensures top talent is not just attracted but also keen to stay and grow.

Finally, Musk has shown the world that failure, under the right circumstances, is an option. He has publicly expressed that if you're not failing, you're not innovating enough. This mindset has created a culture where taking risks is not just allowed but celebrated. This attracts those minds that are tired of playing it safe, those who want to push boundaries, test limits, and create the future.

Renowned tech expert Tim O'Reilly has said, "Musk has essentially turned the Silicon Valley ethos of 'fail fast' into 'fail epically.' That's a powerful draw for talent."

To imagine Musk's strategy is akin to envisioning a formidable sports coach. This coach is not simply seeking to win games; he's seeking to redefine the sport. He pushes his team hard, perhaps harder than they've ever been pushed, but he's right there with them, pushing the boundaries. He doesn't merely accept failure; he celebrates the audacious attempts that lead to it.

As we dig deeper into Musk's method of attracting, nurturing, and retaining talent. It's important to remember that these strategies are not just for CEOs and entrepreneurs but also for anyone aiming to lead, inspire, and achieve remarkable things.

A key part of Musk's strategy is transparency. He maintains open lines of communication and isn't afraid to share the good, the bad, and the ugly with his team. During SpaceX's initial rocket failures, he openly admitted the issues and used them as motivation to improve. The Economist's report on SpaceX in 2012 noted, "Musk's honesty during adversity built a strong trust within his team and instilled a culture of openness and collaboration." This culture attracts those who value truth over illusion, making SpaceX a hub for driven, passionate talent.

Another crucial factor in Musk's talent magnetism is his insistence on continuous learning. This expectation for self-improvement isn't limited to certain roles or departments. It permeates all levels of his organizations. In a famous email to Tesla employees, Musk wrote, "Everyone should be always learning. You should always seek to improve yourself and broaden your perspective." Encouraging learning and development at every level can be a potent tool in fostering employee satisfaction and loyalty.

Moreover, Musk's audacious projects offer the possibility of creating something truly pioneering. He's not looking to make slight improvements or modest advancements; he's after monumental change. This audacity draws those who are keen to leave a mark on the world, those who find satisfaction not in routine but in the thrill of breakthroughs and discoveries.

Think of it like this: Musk is a symphony conductor, his companies are the orchestra, and his ambitious goals are the extraordinary compositions they're attempting to perform. He brings together diverse talent, each a virtuoso in their own right, united under the grand vision of creating music that's never been heard before.

By viewing his workforce not as employees but as critical contributors to his audacious missions, Musk has managed to create an ecosystem that pulls in talent like a strong gravitational force. The harmony he creates between a clear, world-changing mission, a challenging yet supportive work environment, and a culture of transparency and continuous learning, is a blueprint for modern leadership.

This blueprint is not just for industry giants but also for budding entrepreneurs, team leaders, and managers. If you have a team, ask yourself: Are you creating an environment where they feel part of something larger than themselves? Are you challenging them while also providing the necessary support? Are you fostering a culture of openness and continuous learning?

As we conclude this chapter, let us ask ourselves: What does this tell us about the future of leadership? What can we learn from Musk's strategies in our own quests to attract and nurture talent? As we venture into an era of unprecedented innovation and change, it might just be worth considering the 'Musk Method' in our playbook.

In the end, Musk's talent magnetism is more than simply a human resources strategy; it's a core value that informs every part of his businesses. It is the fuel that pushes invention and drives his businesses to heights they could never have imagined. And perhaps we should take a leaf from Musk's playbook as we work to complete tasks, to upset the status quo, and to think broadly. The chances may not be in your favor, but as Elon Musk put it, "When something is important enough, you do it even if the odds are not in your favor."

30. Iterative Excellence: The Pursuit of Quality

The fierce rivalry and constant speed in the field of technological innovation are well recognized. Complacency is a formula for oblivion in this world. Elon Musk's approach to guaranteeing perfection in all of his endeavors can thus be summed up in two words: continual iteration and critical feedback.

Musk's technique includes a component of critical feedback that goes beyond merely accepting constructive comments. In actuality, Musk deliberately seeks it out. According to him, individuals frequently hesitate to own their mistakes, which results in a hazardous lack of awareness. Musk fosters a culture where staff members are allowed to criticize the company's operations, even if it is directed at him, as a preventative measure.

The Wall Street Journal, in its 2015 profile on Musk, reported a case where a junior engineer at SpaceX pointed out a design flaw in a rocket component that Musk himself had proposed. Rather than reacting defensively, Musk thanked him and promptly incorporated the feedback. This incident serves as a testament to Musk's commitment to seeking critical feedback, promoting an open and honest environment where betterment takes precedence over ego.

This ethos was put on full display during the development of Tesla's Model S. while other automakers might have been content with the acclaimed sedan's initial success. Musk pushed his team for improvements, issuing 20 new iterations in the first year of production alone. As Harvard Business Manager reported, this "relentless iteration has become the secret sauce behind Tesla's industry-leading quality and innovation."

Consider yourself the chef in a renowned restaurant to have a better understanding of these guidelines. The constructive criticism is comparable to having a group of dependable tasters who won't hesitate to let you know when a meal requires extra seasoning or a different component. On the other hand, continuous iteration is like perfecting that recipe, making little adjustments each time until the meal is as good as it can be.

But why are these principles so effective? It's simple. They cultivate a culture where complacency is the enemy and improvement are the norm. They foster an environment where criticism is not just expected but embraced, where the status quo is constantly challenged, and where excellence is the target and not an accidental result.

In the larger context of getting things done, these principles ensure that not just any result, but the best possible result is achieved. They ensure that projects aren't just completed but refined and perfected. They are the guardrails that keep Musk's ventures on the path to excellence.

Musk's method of quality assurance underscores an important question for all of us: How open are we to feedback, and how committed are we to continual improvement? As we navigate our own entrepreneurial and professional journeys, it might be worth reflecting on this 'Musk Method.'

In the words of Musk himself, "If you're not failing, you're not innovating enough." How are you integrating failure and feedback into your path to innovation and excellence? How can you foster a culture of critical feedback and constant iteration in your own spheres of influence? These are questions worth pondering as we aim to disrupt, think big, and get things done in our respective fields.

Firstly, it's essential to understand that Musk's approach to feedback is comprehensive and all encompassing. It isn't restricted to a select group of executives or specific stages of a project. Instead, it permeates every level of his companies and is embedded in every phase of production. He creates an environment that promotes candid, fearless communication, where critical feedback is not just welcomed but actively encouraged.

This concept has roots in Musk's belief, as reported in the New York Times, that "One should take the approach that they are wrong. Their goal is to be less wrong." It might seem counter intuitive, but accepting that you could be wrong frees you to listen, learn, and adapt, leading to a more robust final outcome.

The second principle of constant iteration is integral to Musk's mission to create cutting-edge technology. Tech expert Robert Scoble once wrote, "Musk is never satisfied with 'good enough.' He's always looking for ways to make things better, faster, more efficient. It's what makes his companies leaders in innovation."

A stellar example of this philosophy in action is SpaceX's development of reusable rockets. Rather than resting on the laurels of initial successful launches, Musk's team continued to refine and perfect the technology. The iteration didn't stop even after the rockets were successfully landing back on Earth. Each launch, each landing, every minor glitch, is an opportunity for learning and improvement.

Drawing an analogy from nature, Musk's method mirrors the process of evolution. Life on Earth didn't just appear in its present complex, diverse forms. It went through countless iterations, each version refining and adapting to its environment over millions of years. Musk has simply taken this time-tested process of nature and applied it to technological innovation, creating companies that continuously evolve, adapt, and improve.

This culture of continuous improvement inspires workers as well as produces better goods. With each iteration, they enjoy the gratification of watching their work becoming better and the things they contribute to getting better. It encourages a sense of advancement and accomplishment, which intensifies the desire for greatness.

In conclusion, providing constructive criticism and iterating often are not simply techniques; they are mindsets. They serve as proof that there is always an opportunity for development, regardless of how great or accomplished you may become. They represent a way of thinking that sees every setback as a learning opportunity and every success as a stepping stone.

31. The Speed of Execution: Turning Ideas into Reality

"In the world of ideas, speed is paramount," says Elon Musk, an entrepreneur known as much for his groundbreaking visions as for his capacity to bring them to life. This chapter delves into Musk's proficiency in executing ideas quickly and effectively—a characteristic that has become synonymous with his identity as a leader and a disruptor.

Elon Musk's ventures, from SpaceX and Tesla to Neuralink and the Boring Company, all bear testament to his ability to not just conceive transformative ideas but also to put them into action at a speed that leaves many in awe.

In a 2018 piece, The Wall Street Journal addressed how Musk's strategy was unique from that of the rest of the car business. While it may take a typical carmaker five to seven years to construct a vehicle, Tesla, under Musk's direction, was able to release the Model S three years after its idea was revealed. This incredibly quick execution played a crucial role in Tesla's emergence as a significant player in the automobile industry.

At SpaceX, the speed of execution became a lifeline for the company's survival. According to a New York Times feature, in the company's early years, after three failed launches, SpaceX was running out of funds. The fourth launch had to succeed, or the company would fold. Under this pressure, Musk rallied his team to make the necessary modifications and prepare the rocket for launch in a few months. The successful launch marked the turning point for SpaceX, and the rest, as they say, is history.

But how does Musk manage such speedy execution? What is his secret?

The answer lies in Musk's unique approach to problem-solving. He uses what is known as 'First Principles Thinking', a method that involves breaking down complex problems into their basic elements and then reconstructing solutions from the ground up. This approach helps in bypassing traditional bottlenecks and fostering innovative solutions, enabling faster execution.

Consider yourself constructing a house of cards in order to grasp this idea. The traditional method may be to stack cards on top of each other, constructing your construction over time. But if the foundation is weak, the whole structure might fall. Instead, you would examine the surface, balance, and qualities of the cards using First Principles Thinking, and then create a more reliable and effective structure.

By fostering a culture that encourages radical problem-solving methods, Musk has built a framework that accelerates the speed of execution. But it's not just about speed. It's about effective, thoughtful speed, as Musk knows that moving fast without a solid plan or a clear objective can lead to disastrous outcomes.

Moreover, Musk's leadership style substantially adds to this speed of execution, as noted in a Harvard Business Manager report. He takes the initiative, pushing his troops to keep up with his fast speed. His organizations appear to be infused with his unrelenting energy and drive, which encourages everyone to work as quickly and efficiently as possible.

In the context of 'Getting Things Done', speed of execution plays a critical role. In a world that is evolving at an unprecedented pace, speed is the currency of success. It is about keeping up with the relentless march of progress, about making the most of the present while shaping the future.

As we face our own challenges and goals, we should reflect on our speed of execution. Are we turning our ideas into reality swiftly and effectively? Or are we allowing hurdles and apprehensions to slow us down ? Are we ready to run, ready to chase our visions with the same intensity and commitment as Elon Musk?

The inertia of status quo can be a daunting barrier, but as Musk has repeatedly shown, it is a barrier that can be overcome with the right mindset and approach. The fundamental question, then, becomes: Are we satisfied with simply being dreamers, or do we aspire to be doers, individuals who have the courage to shape their ideas into tangible realities?

To use a track and field analogy, ideas are the starting blocks, the springboard from which we launch ourselves towards our goals. Execution, then, is the sprint, the surge of energy and effort that propels us forward. And speed of execution? That's the wind at our back, pushing us to reach our destination faster, ensuring that we don't just finish the race, but that we strive to be among the front runners.

As we forge ahead, let's remember that while speed is a valuable ally, it must always be balanced with accuracy and efficiency. Speed, for the sake of speed, can often lead to errors and oversights. Instead, like Musk, we should aim for thoughtful speed, the kind that prioritizes progress but never at the cost of quality.

Taking a page from Musk, we should infuse our execution process with a feeling of urgency while keeping in mind the bigger picture of our objective. We should develop the ability to deconstruct complicated issues into simpler, more manageable parts before tackling them with bold action and unflinching commitment.

. . .

As this chapter comes to a close, we may draw a crucial conclusion from Musk's lightning-quick execution. It takes guts, leadership, and tenacity to carry out creative ideas in a timely and efficient manner in addition to possessing such ideas. Musk himself stated that "when something is important enough, you do it even if the odds are not in your favor."

Let us then ask ourselves: How quickly and effectively are we turning our ideas into reality? Are we prepared to sprint on the track of our dreams, propelled by the wind of execution? The answers to these questions will shape not just our journeys, but our destinies as well.

As we navigate our paths, let us remember Musk's words, "It's OK to have your eggs in one basket as long as you control what happens to that basket." In the quest to disrupt, think big, and get things done, let's ensure that we're not just controlling our basket of ideas but also actively turning them into the realities of tomorrow.

32. Leading by Example: Musk's Leadership Model

We've spent a lot of time exploring Elon Musk's thoughts, one of the most significant tech entrepreneurs of the twenty-first century. It's time to analyze the fundamentals that have helped him reach such incredible heights: his leadership style, his team-building techniques, and the corporate culture he promotes.

Elon Musk's management approach involves total involvement. Between him and the task, there is none. He is pushing the boundaries of what is thought to be feasible in the trenches with his squads. According to the Wall Street Journal, Musk reportedly slept on the floor of his Tesla plant, exhibiting a level of passion and dedication that is highly valued by his team.

This level of hands-on leadership serves a dual purpose. On one hand, it allows Musk to intimately understand the challenges and obstacles his teams face. On the other hand, it sends a powerful message about the ethos of hard work and dedication that Musk embodies, thereby inspiring his employees to mirror that same passion.

Beyond his leadership style, Musk's approach to team-building is equally remarkable. Unlike traditional hierarchical structures, Musk champions a flat structure, where the free flow of communication and ideas is encouraged. As noted in Harvard Business

Manager, this has led to an organizational culture where every idea no matter how unconventional is up for consideration. Musk knows that in a world that's rapidly changing, it is innovation that holds the key to survival and success.

Musk's team-building approach, though, is about more than simply encouraging innovation—it's also about appreciating the value of human capital. Ben Horowitz, a venture capitalist and technology specialist, said in the New York Times that "Elon believes in people more than anything else. He knows that it is people who drive change, and he has a knack for spotting talent and pushing them to their limits."

This dedication to his teams creates a culture within his companies that is as relentless as Musk himself. It's a culture that doesn't just tolerate failure but rather celebrates it as a step closer to success. At SpaceX, they call this the "Fail Fast" culture. It's the understanding that every failure brings with it a valuable lesson that pushes the team one step closer to their goal.

In fostering this culture, Musk demonstrates another key principle of his leadership: the ability to shift perspectives. Rather than viewing failure as a setback, Musk has conditioned himself and his teams to see it as a stepping stone towards success.

But how does Musk's leadership style translate into tangible results?

Through this immersive leadership, Musk encourages a culture of relentless innovation, a culture that has resulted in groundbreaking achievements. Under his leadership, Tesla has revolutionized the automotive industry, SpaceX is on track to make interplanetary travel a reality, and Neuralink is pushing the boundaries of neurotechnology.

The real magic of Elon Musk, however, lies not just in what he has achieved but how he has achieved it. His leadership is a masterclass in transforming ambitious visions into reality. It's about having the courage to challenge the status quo and to venture into the unknown.

We start to recognize the Musk Model of Leadership when we look at Musk's leadership style, approach to team building, and the culture he fosters. This model is based on a strong commitment to a vision, a firm confidence in people, and an unwavering pursuit of innovation.

In the words of the man himself, "If something is important enough, you should try, even if the probable outcome is failure." This sentiment encapsulates Musk's relentless pursuit of progress, despite the odds.

Musk approaches "getting things done" with a certain amount of audacity, an unwillingness to accept "good enough," and a firm conviction that failure is not the antithesis of success but rather a step on the path to it. It's a leadership style that aims to inspire people to develop their skills and fan the flames of creativity rather than just manage them.

What does this indicate about the potential influence of the Musk Model on next entrepreneurial and innovative generations? It implies a fundamental change in the way we think about business innovation, favoring audacity over prudence, unrelenting optimism over skepticism, and faith in people over institutions and processes.

As futurist and renowned tech expert Amy Webb once said in The Economist, "Musk's leadership style isn't just a model, it's a mindset. It's a challenge to the status quo. And whether you agree with him or not, there's no denying that it's a mindset that's shaping our future."

In a world constantly being redefined by technology, Elon Musk stands as a beacon of relentless pursuit of innovation. His model of leadership is not merely a set of principles but a blueprint for 'getting things done.' It's an approach that's pushing the boundaries of what we believe to be possible, shattering conventional norms, and opening doors to a future that, under any other circumstances, would seem like the stuff of science fiction.

As we wrap out this chapter, it's important to consider Musk's ethos and how it pushes us to reevaluate how we lead and accomplish our goals. It serves as a reminder that the future is something that we actively choose. Instead, it's something we actively influence by today's choices and behaviors.

Musk's leadership demonstrates that sometimes, to create something groundbreaking, we must be willing to take risks, to challenge the established norms, and most importantly, to believe in the power of our own visions. His journey is a testament to the adage, "What you can conceive and believe, you can achieve."

At the intersection of technology and humanity, Elon Musk stands as a visionary force, leading by example, and relentlessly pushing the envelope of innovation. His journey serves as a testament to the transformative power of disruptive thinking and daring leadership.

33. The Future is Now: Transforming Visions into Reality

The universe of Elon Musk is an assembly line of dreams, moving from concepts to reality. It is in this transformation where Musk truly excels, demonstrating a remarkable capacity for getting things done. Musk's audacious objectives have always been loftier than mere business success. He seeks to redefine our world fundamentally and our place in the cosmos. But how does he convert these towering ideas into concrete reality?

Musk adheres to a unique set of principles, honed by years of experience, relentless curiosity, and an unwavering commitment to his visions. The core of these principles is his 'First Principles Thinking,' a concept borrowed from physics but effectively applied to entrepreneurship. This thinking involves breaking down complex problems into their fundamental elements and then reassembling them from the ground up. For Musk, this approach isn't just a theoretical exercise; it's a pragmatic tool that has aided him in designing groundbreaking electric vehicles and reusable rockets.

Consider Tesla's disruptive innovation, the electric car's battery. The conventional wisdom was that batteries were too costly to make electric cars viable. But Musk questioned this assumption, analyzed the individual components of a battery, and realized that by redesigning and manufacturing them in-house, he could dramatically reduce costs. This first principles approach allowed Musk to achieve what many had deemed impossible.

Musk's second principle is to welcome failure as a learning opportunity. Whether it was PayPal's early struggles, SpaceX's launch failures, or Tesla's near bankruptcy, Musk has always used setbacks to iterate, improve, and advance. His perspective reframes failure from being an endpoint to a necessary step on the road to eventual success.

Accompanying his acceptance of failure is Musk's commitment to a rigorous feedback loop. He constantly seeks out critique especially on his shortcomings. He regularly consults with experts, employees, and even his customers. By fostering an environment that encourages open communication, Musk ensures that he and his teams are continually learning and evolving.

Lastly, Musk embodies an unyielding perseverance, often referred to as the "Musk effect." It is a potent mix of optimism, self-confidence, and tenacity. He sets seemingly impossible goals and pursues them relentlessly, often working hundred-hour weeks. His deep-rooted belief in his missions' validity propels him and his teams to continually innovate, iterate, and improve.

Musk's approach is neither straightforward nor easy. It requires deep thinking, a willingness to learn from failure, open-mindedness, and unwavering determination. But as SpaceX's reusable rockets ascend into the cosmos and Tesla's electric cars traverse our roads, the success of Musk's methods is undeniable.

In an era of incremental innovations, Musk's ambitious ventures remind us of the power of bold thinking and steadfast execution. His entrepreneurial journey sends a compelling message to us all: dream big, question conventions, and don't shy away from taking calculated risks.

Now, imagine for a moment what our world could be if we all adopted some measure of Musk's audacity, determination, and unwavering belief in the future. Could you, too, transform your visions into reality? As we journey forward, remember: the future, in the realm of Elon Musk, is always now.

One cannot understate Musk's role as a visionary leader in getting things done. He isn't just directing the course; he's part of the crew, immersed in the details and intricacies of his projects. This is best evidenced by Musk's dual roles as CEO and lead designer at SpaceX. It might seem improbable for one of the world's busiest CEOs to also be an engineer, but for Musk, it's his modus operandi.

Consider his decision to learn rocket science. When Musk first dreamed of space exploration, he had little background in the field. Nevertheless, he devoured books, consulted experts, and taught himself the principles of rocketry. This self-directed learning approach enabled him to lead SpaceX into becoming a pioneering force in commercial space travel.

While most CEOs might step back once the business model is established, Musk dives deeper. He is known to spend significant time at Tesla's factory floors, working with engineers, understanding the bottlenecks, and troubleshooting problems. His hands-on approach sends a powerful message to his teams about his commitment and work ethic, inspiring them to follow suit.

Musk's work culture is fueled by a strong sense of urgency. For him, time is the most valuable commodity, and he expects his employees to share this view. He encourages rapid iteration and swift decision-making. This "move fast and fix things" mindset enables Musk's companies to innovate at a pace that leaves many competitors scrambling to keep up.

Moreover, Musk's vision transcends the balance sheets. His ultimate goals – sustainable energy and interplanetary life – aren't merely about profit. They're about ensuring humanity's survival and prosperity. This sense of purpose imbues his work and his companies with a powerful drive that motivates employees and attracts a dedicated customer base.

It's essential to remember that Musk's methods might not suit everyone. His style is intense, his expectations high, and his visions, often, out of this world. Yet, the efficacy of his approach cannot be dismissed. Through Tesla, SpaceX, Neuralink, and SolarCity, Musk has shown that a daring vision, combined with relentless execution, can transform industries and change the world.

As this chapter draws to a close, we return to the core of Elon Musk's brilliance: the capacity to imagine large, think creatively, and make those goals come true. He does this not via magic or just charismatic means, but rather through careful consideration, ongoing education, unrelenting effort, and a business confidence in a better future.

In the words of Peter Drucker, "The best way to predict the future is to create it." Musk's entrepreneurial journey gives life to this quote, showing us how a single individual, armed with the right mindset and principles, can indeed shape the future. It is a testament to the power of disruptive thinking, audacious goals, and determined execution.

Keep in mind that Elon Musk started his career as an entrepreneur just like anybody else when you put this book away and resume your regular activities. His capacity to dream large, accept failure, and take on apparently impossible undertakings sets him unique.

Afterword

As we reach the end of this journey through Elon Musk's extraordinary life and groundbreaking ventures, it's crucial to pause and reflect on what we've uncovered. This has been more than just an exploration of a man's life; it's been an illuminating guide on how to disrupt, think big, and get things done.

The lessons we've learned from Elon Musk's story are invaluable. He teaches us that no idea is too ambitious, no vision too grand, and no obstacle too daunting. He shows us that innovation isn't merely about creating something new, but about daring to reimagine and redefine what's possible.

Yet, perhaps the most powerful takeaway is the understanding that each of us has the potential to impact the world. Musk's journey reveals that with courage, determination, and a relentless belief in our dreams, we too can spearhead change and transform the seemingly impossible into reality.

In the spirit of Elon Musk, let's carry these lessons forward. Let's dare to dream bigger, push harder, and strive for greater heights. Above all, let's remember that our capacity to shape the future is only as limited as our courage to disrupt, our ability to think big, and our commitment to getting things done.

As we close this book, remember that the real adventure is just beginning. Armed with the lessons we've learned, it's now our turn to step into the arena of disruption, and, like Musk, leave our own indelible marks on the world. The future awaits.

Made in the USA
Middletown, DE
13 December 2023